A FIELD EXCURSION GUIDE TO THE

ISLAND OF MALLORCA

by

H. C. JENKYNS, B. W. SELLWOOD & L. POMAR

Edited by C. J. Lister

©THE GEOLOGISTS' ASSOCIATION
1990

AUTHORS:

Hugh C. Jenkyns,
Department of Earth Sciences,
University of Oxford,
Parks Road,
OXFORD, OX1 3PR.

Bruce W. Sellwood,
The Postgraduate Institute for Sedimentology,
University of Reading,
Whiteknights,
READING RG6 2AB.

and

Lluis Pomar
Departament de Ciències de la Terra,
Universitat de les Illes Balears,
Carretera de Valldemossa, km 7.5,
07071 Palma de Mallorca,
SPAIN.

Other contributors:

Elaine Jones (Tertiary sediments),
Shell International Petroleum Mij. B.V.,
The Hague,
NETHERLANDS.

David Prescott (Jurassic-Cretaceous sediments),
Department of Earth Sciences,
University of Oxford,
Parks Road,
OXFORD OX1 3PR.

John Senior (ammonites),
Department of Extramural Studies,
University of Durham,
32 Old Elvet,
DURHAM DH1 3HN.

Antonio Rodríguez-Perea (regional geology)
Emilio Ramos (Tertiary sediments)
Departament de Ciències de la Terra,
Universitat de les Illes Balears,
Carretera de Valldemossa, km 7.5,
07071 Palma de Mallorca,
SPAIN.

Francesc Sàbat (structural geology)
Departament de Geología Dinàmica,
Geofísica i Paleontología,
Universitat de Barcelona,
Zona Universitària de Pedralbes,
08028 Barcelona,
SPAIN.

(ii)

CONTENTS

List of Figures

(iv)

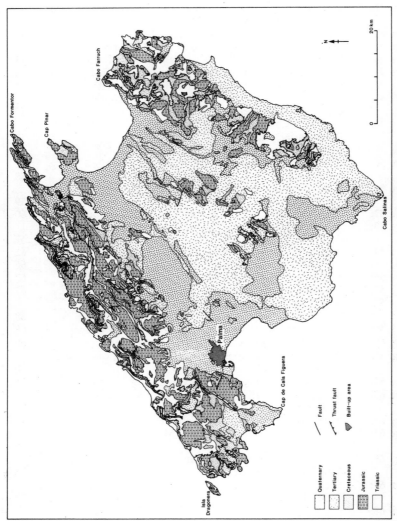

Fig. 1. Simplified geological map of Mallorca, after Mapa Geológico, Palma de Mallorca-Cabrera, sheet 57–66, 1:200,000, Instituto Geológico y Minero de España (1970).

(vi)

PREFACE

This guide is based on numerous trips we have led to the island of Mallorca in the last decade, including two for the Geologists' Association, and contains a certain amount of original research. Data have also been drawn from a number of sources. We wish particularly to acknowledge the Excursion Guide to the Tertiary of the Balearics produced by the Spanish Sedimentological Group, the Sedimentology of the Jurassic of Mallorca produced by the Spanish Mesozoic Group and the Ph.D. theses of Elaine Jones, David Prescott, Emilio Ramos, Antonio Rodríguez-Perea and Francesc Sàbat. We are also grateful to all those who accompanied us on trips to the island, collected fossils, pointed out features in the rocks that we ourselves had missed, and examined various drafts of the manuscript; we acknowledge particularly the help of Trevor Burchette, Jim Kennedy, John Platt and Maurice Tucker. A financial subvention from British Petroleum, which aided in production of the initial version of this guide, is also recorded with thanks.

In terms of logistics it is assumed that people will be based in the area close to Palma. The guide is primarily designed to be used by those travelling in their own or hired cars: a few of the suggested routes, utilizing minor roads (particularly Localities 11 and 18), cannot be traversed by coach. The programme for each day is full and if the trip is taken in the winter months there will probably not be sufficient light to visit all the recommended exposures, if the party is large and/or slow-moving. Optional localities have been included during several days and these can be abandoned if time is short. Those seeking a 5- rather than 6-day excursion are advised to dispense with Excursion 3. Note that the spelling of place names varies somewhat according to whether they are written in Mallorquin or Spanish: there has been no great effort to strive for consistency in this guide. Note also that Mallorca is undergoing continuous development, particularly in the resort areas close to Palma and the locality details may become outmoded with time. Parts of Mallorca are still *terra incognita* geologically speaking; witness the discovery of Palaeozoic rocks which were unknown a few years ago. Geological research has been particularly active in the area since 1980 and continues apace, so many concepts and interpretations will change.

Included here are two figures, one (Fig. 1) representing a simplified geology of the island, and another (Fig. 2) showing the location of the stops. The geological map of the area (1:200,000) is published by the Instituto Geológico y Minero de España: 57 – 66, Mallorca-Cabrera. Most of the road maps on sale in the island (e.g. Baleares 1:125,000, Firestone Hispania) are adequate for the needs of this excursion.

It is important to stress that geological studies on Spanish territory require permission from the Geological Commission in Madrid. This can be obtained by writing to: Comisión Nacional de Geologia, Rios Rosas 23, MADRID-3, SPAIN. The granting of such permission for a field excursion is normally a formality.

1

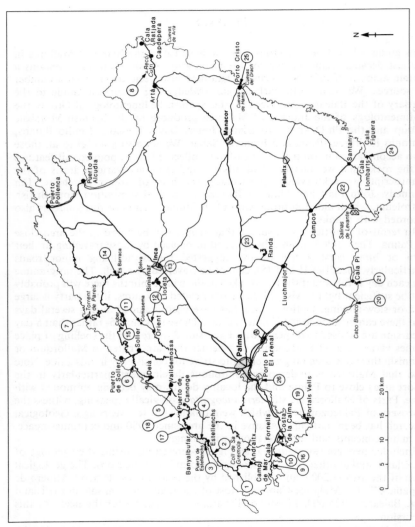

Fig. 2. Map showing localities described in this guide.

GEOLOGY OF THE BALEARIC ISLANDS, WITH PARTICULAR REFERENCE TO MALLORCA

INTRODUCTION

The Balearic Group, the most isolated of the Mediterranean islands, covers an area of 5014km². Its submarine platform is separated from the east coast of Spain by the Valencia Trough, a channel over 1000m in depth, but it continues the La Nao promontory to form a ridge, the 'Promontorio Balear', essentially a submarine continuation of the Betic Cordillera. The alignment of islands runs approximately east-west along a 300km belt (Fig. 3). The largest island is Mallorca (Majorca) and covers 73% of the area. Following in order by size come Menorca (Minorca), Ibiza (Eivissa, Iviza), Formentera, and Cabrera. Ibiza lies about 100km off the Valencian coast.

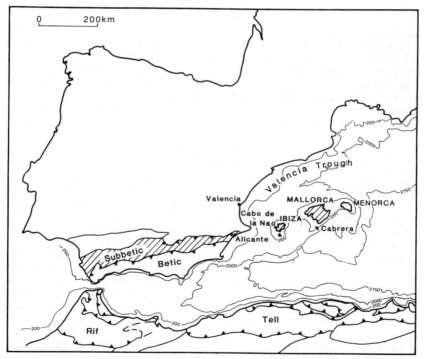

Fig. 3. Mallorca and the other Balearic Islands in their geotectonic context in the western Mediterranean. After Bourgois *et al.* (1970), Azéma *et al.* (1974) and Bernoulli and Jenkyns (1974). Mallorca and Ibiza are conventionally assigned to the Subbetic Zone as the Mesozoic facies belts are closely comparable. Submarine contours in metres.

3

The islands enjoy a typically Mediterranean climate controlled by the dynamics of the westerlies and their Mediterranean branch with a characteristic air-mass; accordingly the annual precipitation – most of it in spring and autumn – varies between 300mm in the southern plains and 1400mm in the mountains; such amounts do not provide for the continuous flow of surface waters. The yearly temperature average of 17°C is based on mild winters (10°C in January) and hot and dry summers (24°C in July). Regarding the soils, there is a variety of types such as rendzinas, xero-rendzinas and humid rendzinas and terra fuscas in the mountains; whereas in the plains, terra rossa sediments are widespread; and serozem occurs in the arid plateaus of Ibiza and Formentera and locally in Mallorca. The terra rossa soils, assumed to derive largely from Saharan air-borne dust, are relict (Crabtree et al., 1978). Where man-made erosion has been relatively mild, these soils remain covered by the native vegetation which includes many forms endemic to the islands, particularly at higher altitudes (Colom, 1957).

MORPHOLOGY AND STRUCTURE

Structurally, the islands trend north-east – south-west, representing a prolongation of the alignments of the Subbetic Cordilleras of southern Spain (Fig. 3) with imbricate structure along a sharp front on the north-west sides of Ibiza and Mallorca; the tectonic connection with Menorca and the rest of the Alpine Zone is cut off by major faults. Seismic data suggest that the Mesozoic and Cenozoic sedimentary cover in Mallorca is some 7km thick, whereas in Ibiza it is only 4km (Banda et al., 1980). The Moho lies at a depth of about 20km under Ibiza, increases to 25km under Mallorca and rises to 18km under Menorca.

Mallorca (Majorca)

The island of Mallorca covers 3640km^2 (96×78km); it displays three major morphotectonic divisions (Fig. 4). The Sierra Norte (Serra de Tramuntana) forms the north-west side and comprises several tectonic zones (Fallot, 1922), mainly built of Mesozoic sediments, cut by longitudinal valleys, karstic canyons and poljes, and with a spectacular 'costa brava' (cliffed coast). Fallot distinguished a lower unit with red Triassic Buntsandstein exposed on and above the north-west coast and including a small outcrop of recently discovered Palaeozoic rocks (Ramos and Rodríguez-Perea, 1985), a second unit with the highest mountains including Puig Major (1445m), and a more southerly disposed unit. The more recent scheme of Alvaro et al., 1984, identifying five tectonic units in the Sierra Norte, is illustrated in Fig. 5. Major thrusting from the south-east took place during early Miocene (Langhian) time (Colom, 1975b; Pomar, 1979; Boutet et al., 1982) although there is local evidence for prior (pre-Burdigalian) movements also (Pomar et al., 1983a).

There are numerous cold-climate Pleistocene phenomena at high altitude here (over 900m); frost-shattering and scree formation is currently significant only on Puig Major where the mean January temperature is −1°C (Crabtree et al., 1978). Further down the slopes, colluvial deposits and terra rossa palaeosols are important (Butzer, 1962, 1964).

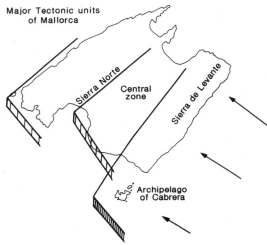

Fig. 4. Major threefold tectonic divisions of Mallorca, based on Colom (1950, 1973). Recent workers assume Cabrera to be structurally integral with the Sierra de Levante (Sàbat and Santanach, 1985).

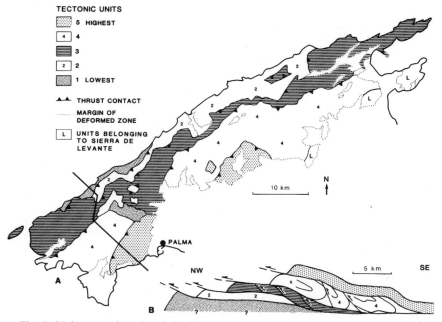

Fig. 5. Major tectonic units of the Sierra Norte, after Alvaro *et al.* (1984) and Prescott (1989). A, structural map with line of section; B, cross-section.

The central zone between the two major ranges is a relatively low plain in which up to five geomorphological divisions may be distinguished (Fig. 6):

1.   Es Pla, principally comprising deformed rocks of Mesozoic to Oligo-Miocene age.

2.   Es Raiguer, including the Palma, Inca and Sa Pobla Basins which subsided during Late Miocene-Pleistocene time.

3.   Campos Basin, subsident during the middle Pleistocene.

4.   Lluchmajor Marina characterized by a platform produced by progradation of a Tortonian-Messinian reef across a deformed Mesozoic-Oligocene substratum.

5.   Santanyi Marina, characterized by a reefal platform and post-reef sediments which encircle the Sierra de Levante.

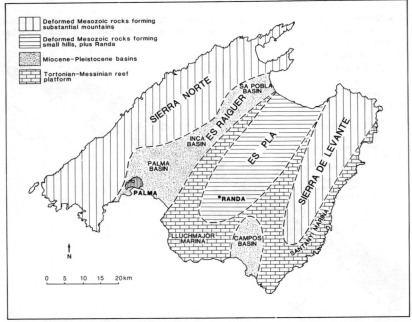

Fig. 6. Major tectono-sedimentary domains of Mallorca. Modified after Pomar *et al.* (1985).

Some of the central-zone hills (Puig de Randa) have proved tectonically problematic: the observations of largely undeformed Burdigalian resting discordantly on near-vertical marine Oligocene (Colom, 1975a; Colom and

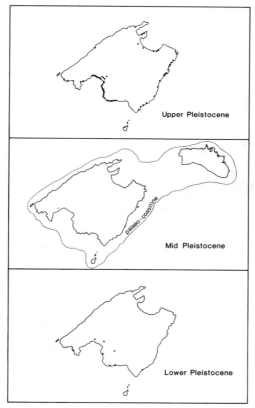

Fig. 7. Distribution of Lower and Upper Pleistocene sediments on Mallorca. Note the more inland positions of some Lower Pleistocene deposits; Upper Pleistocene deposits are concentrated on the south-western coast of the Central Zone. The Mid-Pleistocene sketch shows possible position of the shoreline during the Riss glacial maximum. Modified after Cuerda (1975).

Sacares, 1976) suggested possible pre-Miocene thrusting; however, reinterpretation of this area has shown that the Oligocene is allochthonous and the thrusting is Langhian in age (Anglada *et al.*, 1986).

Lower Pleistocene sediments are known from the south-western coast and inland (Fig. 7). On the same coast there are also well-preserved and widely distributed marine sediments of late Pleistocene age (Cuerda, 1975). Raised marine platforms and notches attest to former high stands of sea level (Fig. 8). During the early stages of each regression, coastal dune sands accumulated and were subsequently lithified into aeolianites. They are interstratified with loessic palaeosols and weathered to terra rossa and calcrete duricrusts (Butzer and Cuerda, 1962).

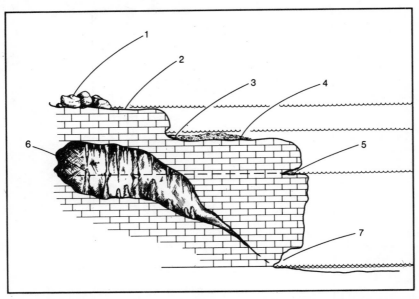

Fig. 8. Sketch showing Pleistocene phenomena commonly encountered where Miocene carbonates are exposed on the coast, as they are in parts of the Central Zone and east of the Sierra de Levante. 1=boulders of rounded Miocene carbonate; 2=platform produced by marine abrasion representing the highest former stand of sea level; 3=wave-planated surface similar to 2; 4=relict beach sand and conglomerate; 5=erosional/solutional notch produced by former sea-level stand; 6=cave, decorated with stalagmites and stalactites, assumed to have been initiated when sea level stood at the datum indicated by 5. 7=marine cavern currently being produced by erosion and dissolution. After Cuerda (1975).

The Sierra de Levante (Serres de Llevant) in the southeast constitutes a series of low mountains and hills mainly built of Jurassic limestones and dolomites. Darder distinguished five major tectonic units emplaced, like those in the Sierra Norte, from the south-east (Darder and Fallot, 1926). The more recent studies of Sàbat (1986), Pares et al. (1986) and Sàbat et al. (1988) similarly identified five major structural units in the northern half of the area, and two, structurally higher units in the southern area (Fig. 9). The major thrust movements here were probably initiated during the Oligocene and continued at least until the Langhian (Colom, 1975a; Sàbat, 1986). At the eastern foot of the chain lies Santanyi Marina, which connects southwards with the geologically similar Lluchmajor Marina (Fig. 6). Here are the gorges known as calas, which are structural or gradational subaerially eroded valleys that have been invaded by the sea as rias. Here, also, the shoreline possesses a well-preserved set of upper Pleistocene sea-level records (Fig. 7; Cuerda, 1975).

A structural cross-section of Mallorca is given in Fig. 10.

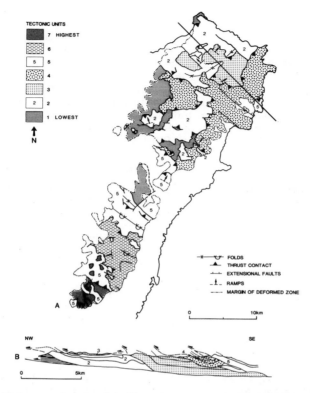

Fig. 9. Major tectonic units of the Sierra de Levante. Modified from Sàbat *et al.* (1988). A, structural map with line of section; B, cross-section.

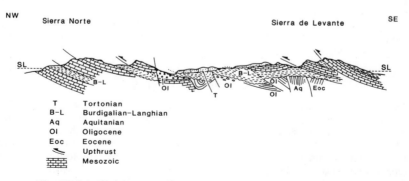

Fig. 10. Schematic cross-section of Mallorca, after Bizon *et al.* (1978).

Menorca (Minorca)

With an area of 702km², Menorca is a similar although more compact island, more nearly east-west oriented, but geologically rather different from Mallorca. In Menorca, Palaeozoic as well as Triassic rocks occur in the northern half (Tramuntana) of the island, opposite the Migjorn, a plateau like the 'Miocene Marina' of Mallorca. Hills and ravines diversify the island, which rises to a maximum height of 358m in the north. Here, too, there are calas with a similar history to those in Mallorca, although the harbours (*puertos*) correspond to the major fault zones, for example that of Maó, a true ria 4 or 5km in length. The shoreline also shows evidence of several Pleistocene sea-level stands.

Ibiza (Ivisa, Eivissa)

Geologically, the island of Ibiza (541km²) is not dissimilar to Mallorca, although its Mesozoic facies are somewhat different. Three nappe units have been recognized by Rangheard (1971) and the island embraces a stratigraphy from Middle Trias (Muschelkalk) to Miocene. Between the above-mentioned thrust units there are valleys and buried plains covered by Helvetian-Tortonian deposits of molasse type.

Formentera.

Formed entirely of Upper Miocene horizontally bedded coral-reef limestones and Plio-Pleistocene aeolianites, Formentera (82km²) is the most arid and the most southerly of the islands. The predominance of limestone has dictated a very well-marked karst with poljes and caves.

Cabrera

The smallest of these islands, Cabrera, is situated south of Mallorca. Here, Jurassic and Cretaceous rocks are covered unconformably by Eocene sediments, localised in graben, and developed primarily in the northern half of the island. Although assigned to a separate tectonic unit by Colom (1973), Cabrera has been recently divided into two structural units by Sàbat and Santanach (1985) and related directly to the south-western part of the Sierra de Levante (Fig. 4). Cabrera continues in the submarine ridge known as the Emile Baudot Bank (minus 92m), which brings the Balearic Platform to less than 100km from the Algerian side of the Mediterranean (Mauffret et al., 1972).

GEOLOGICAL HISTORY

The structural foundations of the Balearic Islands were probably established during the Caledonian orogeny. Pre-Devonian black schistose pelites, chlorite and sericite schists have been dredged from off Menorca (Bourrouilh and Gorsline, 1979). On the island of Menorca itself the Lower Devonian is exposed as further black shales containing thin siliciclastic turbidites; three graptolite zones are represented (Lochkovian of the Bohemian sequence, i.e. Gedinnian and Siegenian). The Middle and Upper Devonian are similarly represented by a flysch facies, locally bearing conodonts and *Tentaculites* (Eifelian-Givetian-Frasnian). In the Upper Devonian, there are increasingly abundant coarse breccias and turbidites ('microconglomerates', formerly interpreted as tillite) with rolled brachiopods, corals, goniatites, and trilobites (Clauss, 1956; Schwarzbach, 1958; Bourrouilh, 1967). There are also slump structures and intercalations of volcanics with sills and dykes (andesite, basalt, and diorite), which continue into the Lower Carboniferous.

The Devonian flysch of Menorca is followed by a Lower Carboniferous radiolarite, capped by shaly and carbonate facies, overlain in turn by a coarsely detrital 'Culm' unit containing abundant turbidites, sand- and mud-flow deposits in which foraminifera of Namurian age and trace fossils are recorded. Multiple Hercynian orogenic phases were succeeded by extensive erosion and a marked hiatus. Post-orogenic Permian is believed to be represented by some unfossiliferous clastics (comparable to the Rotliegendes of northern Europe), a red conglomerate, followed by sandstone (Zechstein) and Kupferschiefer-type shale (Hermite, 1879; Hollister, 1934; Bourrouilh, 1970). The presence of Palaeozoic rocks, formerly thought to be absent on Mallorca, and the difference in structural grain, has led several authors to discuss the so-called 'problem of Menorca' (e.g. Fallot, 1923; Colom, 1950; Bourrouilh, 1970).

The oldest rocks on Mallorca are the recently discovered vertically dipping dark shales and sandstones that outcrop locally on the north-west coast below the Sierra Norte (Rodríguez-Perea and Ramos, 1984; Ramos and Rodríguez-Perea, 1985; see Fig. 11). Although no diagnostic macrofossils have been recovered the

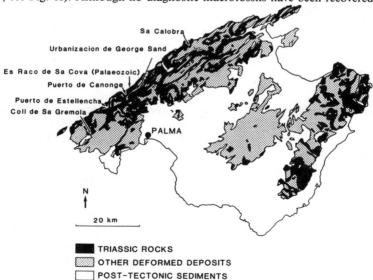

Fig. 11. Distribution of Triassic (and Palaeozoic) facies on Mallorca. After Mapa Geológico, Palma de Mallorca-Cabrera, sheet 57–66, 1:200,000 and Prescott (1989). Key localities indicated.

close lithological similarity with Palaeozoic rocks on Menorca suggests a Late Devonian or Carboniferous age: unpublished palynological data indicate the latter. Permian continental sandstones have been claimed to occur in Mallorca by Colom (1975b) but there is no positive age assignment and they are not mapped by the geological survey; their presence is doubtful. The oldest Mesozoic formations are Triassic, best developed in north-west Mallorca, around Estellenchs, in the mainly continental 'Germanic facies' of red beds and evaporites, with typical tabular cross-bedded fluvial red Bunter sandstones locally containing Lower Triassic plants and remains of reptiles and insects (Rodríguez-Perea et al., 1987; Calafat, 1986; Calafat et al., 1986/87). The distribution of Triassic facies is given in Fig. 11. Bunter facies have been dated as Werfenian in the Betic Cordillera of mainland Spain (García-Hernández et al., 1980).

Above the Buntsandstein follows the calcareous mid-Triassic Muschelkalk locally containing marine fossils such as ceratitids, gastropods, conodonts, bivalves and calcareous algae, in limestones and dolomites (Freeman, 1972), and then red-brown and green gypsiferous marly Keuper, which lacks fossils. Interbedded with these latter are some basic alkaline volcanics, which may be interpreted as related to the initial extensional regime that ultimately led to the rifting of the then-continental western Mediterranean region (Iberian Peninsula, France, Sardinia, and Balearic Islands) as the Tethyan Ocean developed (Bosellini and Hsü, 1973). They include pillow lavas similar to those found in Aquitaine and Morocco. Above the Keuper proper, a unit of grey dolomites and interbedded marls is recognized locally: these may be termed the supra-Keuper and on Mallorca they have yielded palynomorphs of Norian age (Boutet et al., 1982). Similar Germanic-type Triassic facies occur on Menorca and Ibiza, although the Buntsandstein does not outcrop on Ibiza and the Menorcan Keuper contains negligible gypsum and basic volcanics (Hollister, 1934; Azéma et al., 1974; Rodríguez-Perea et al., 1987). A stratigraphic scheme for the Mallorcan Triassic is given in Table 1.

TRIASSIC STAGES	FORMATIONS ON MALLORCA	PRINCIPAL LITHOLOGIES
	Supra-Keuper	marine dolomites
Norian		
	Keuper	?marine to lacustrine marls and evaporites
Carnian Ladinian Anisian	Muschelkalk	marine limestones and dolomites
Scythian/Werfenian	Buntsandstein	fluvio-lacustrine sandstones and siltstones

Table 1. Suggested stratigraphy of the Mallorcan Triassic, after Boutet et al. (1982) and Rodríguez-Perea et al. (1987). Dating is inexact and problematic owing to the lack of diagnostic fossils at most levels.

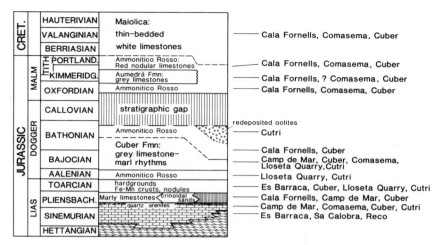

Fig. 12. Generalized Jurassic and Cretaceous stratigraphic scheme based primarily on the work of Alvaro *et al.* (1984) and our own data. This scheme is developed principally for the Sierra Norte but also includes the redeposited oolites of Cutri in the Sierra de Levante. It is still tentative and will undoubtedly be modified by further research.

Lower Jurassic sediments of carbonate-platform facies follow. In Mallorca these are estimated as some 500 metres thick, of which the basal 100-150 metres, attributed to the Lower Lias, are dolomitic with cavernous weathering (carniolas). These facies include oolitic, stromatolitic and pelletal fabrics and contain abundant calcareous algae and foraminifera attributed to the Hettangian, Sinemurian and Pliensbachian (Colom, 1970, 1975b). Intraformational breccias are common (Rodríguez-Perea and Fornós, 1986). The summit of this facies, in both the Sierra Norte and Sierra de Levante, is rich in rounded granules and grains of quartz, providing a distinctive arenitic marker horizon (Fig. 12). These clastics presumably derived from a locally uplifted segment of Triassic Buntsandstein or Devonian-Carboniferous basement.

The break-up unconformity, recognized by an abrupt change to pelagic facies, is dated as close to the Pliensbachian-Toarcian boundary. The oldest pelagic sediments are Pliensbachian, as dated by ammonites and brachiopods locally in the Sierra Norte, close to Soller, and on the island of Cabrera, but Toarcian open-marine facies are widespread over much of the island. In Ibiza, the oldest pelagic facies are Oxfordian red nodular limestones, directly overlying, with considerable stratigraphic gap, lower to mid-Liassic dolomitized platform carbonates which do not exceed 120 metres in thickness (Rangheard, 1971; Azema *et al.*, 1979). Menorca lacks pelagic facies of this age; indeed the Jurassic is entirely represented by shallow-water limestone and dolomites (Azema *et al.*, 1974; Obrador and Fontboté, 1983).

The evolution from shallow- to deep-water environments signified a change from an environment similar to the Bahama Banks to one similar to the Blake Plateau, and reflected major changes in tectonic and/or eustatic patterns

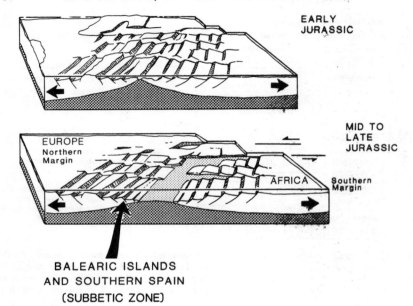

Fig. 13. Idealized scheme showing the evolution of the Tethys during Jurassic time, after
Lemoine (1984). The Subbetic and Balearic Mesozoic successions, with their typical
Germanic Triassic, are characteristic of the northern margin. The Mesozoic of the
southern margin is today represented in the Southern Alps, Eastern Alps, Apennines,
Dinarides and Hellenides (Laubscher and Bernoulli, 1977).

(Bernoulli and Jenkyns, 1974; Schlager, 1981): this was probably the main rifting
event in the Tethys, which was later followed by drifting with a strong lateral shear
component. The Subbetic Zone of Spain and the Balearic Islands became part of
the northern continental margin of Tethys (Fig. 13). The pattern of fault-block
topography, which was probably rather complex, controlled the development of
characteristic sediments (Fig. 14). Mid to Upper Jurassic facies include several
levels of the red nodular limestone known as *ammonitico rosso*, locally slumped
and retextured, and grey limestone-marl rhythms (Bajocian-Bathonian and
Oxfordian-Kimmeridgian) that locally contain replacement chert (Fig. 12). A
stratigraphic gap in the Callovian reflects a Tethys-wide event (Ogg *et al.*, 1983)
and may relate to changing bottom-current systems as Africa began to drift away
from North America to produce the Central Atlantic.

In the Sierra Norte of Mallorca the Tithonian is characteristically marked by
a red-purple *ammonitico rosso* facies. In parts of the Sierra de Levante the Middle
Jurassic (Bathonian) contains redeposited beds of channellised oolite intercalated
within pelagic facies (Alvaro *et al.*, 1984; Fornós *et al*; 1984). The source of this
oolite may be a mid to Late Jurassic Mallorcan carbonate platform (Fornós *et al.*,
1988). Oolitic facies also occur in the Prebetic and parts of the Subbetic Zone of
mainland Spain during this interval (García-Hernández *et al.*, 1979); some of

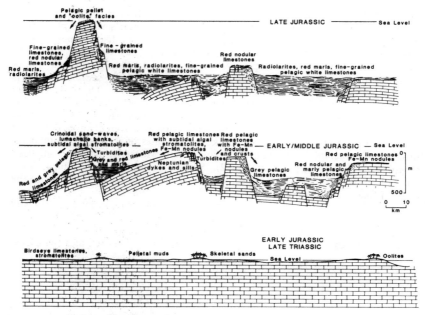

Fig. 14. Generalized scheme, after Bernoulli and Jenkyns (1974), for the relationship between palaeotopography and facies on the northern continental margin of Tethys, based on the stratigraphy of the Subbetic Zone (Seyfried, 1978, 1980; García-Hernández et al., 1980; Comas et al., 1981). Lime-free ribbon radiolarites are not recorded from Mallorca.

these units may also be redeposited as they are intercalated between pelagic sediments. Higher Jurassic sediments in the Sierra de Levante include marls and cherty limestones, locally containing redeposited pelagic and carbonate-platform material.

Cretaceous deposits in Mallorca begin with the *Maiolica*, a white nannofossil limestone containing microfossils (Ciliophora) known as calpionellids; this facies, locally at least, extends down into the Tithonian (Fig. 12). This is a widespread Tethyan sediment known to occur in Cuba, Atlantic deep-sea drilling holes, ubiquitously in the Alpine-Mediterranean region, and the East Indies (Colom, 1955, 1965, 1967a; Bernoulli and Jenkyns, 1974; Ogg et al., 1983). Its development may reflect a rapid increase in the abundance of calcareous nannofossils. In Ibiza similar facies occur in the south-east of the island, but in some areas they contain substantial quantities of derived shallow-water grains; shallow-water carbonate facies occur to the north-west (Rangheard, 1971). Roughly similar facies are known from Menorca (Azéma et al., 1974).

Maiolica facies characterize the interval Berriasian to Lower Barremian in Mallorca and, apart from calpionellids, contain common ammonites and aptychi and are thus well zoned and well dated. The keyhole brachiopod *Pygope* occurs in this lithology in Mallorca and Ibiza (Colom, 1973). Higher sediments (Upper Barremian to Aptian) are generally developed as grey marls with a planktonic microfauna throughout most of the Balearic islands. Rudistid limestones of Aptian age occur in the north of Ibiza, where they may be redeposited. On Mallorca the Albian is developed as organic-carbon-rich blue-black shales containing pyritized ammonites. Glauconite occurs locally in sediments of this age in Mallorca and Ibiza (Rangheard, 1971; Colom, 1984). These facies may reflect the influence of a so-called 'Oceanic Anoxic Event' (= a time of regional oxygen depletion in marine waters) that is registered in many Tethyan areas (Jenkyns, 1980). Upper Cretaceous sediments, first discovered on Mallorca by Colom (1969), comprise white to pink marly limestones: the Cenomanian, Turonian, Coniacian, Santonian, Campanian and Maastrichtian have all been recognized (Alvaro-Lopez *et al.*, 1982). Similar facies occur on Ibiza. Varicoloured marls and marly limestones, with some detrital quartz, characterize the Aptian-Albian of Menorca (Obrador and Fontboté, 1983).

By Early Tertiary time the Balearic region was probably continental, affected by regional orogenic uplift without major deformation of the Mesozoic substratum. The distribution of Tertiary sediments on the island is shown in Fig. 15. No Palaeocene sediments have been recorded from Mallorca and the oldest

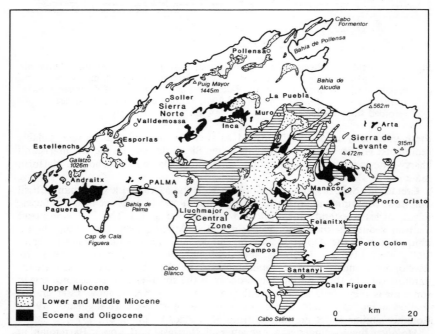

Fig. 15. Distribution of Tertiary sediments in Mallorca. After Mapa Geológico, Palma de Mallorca-Cabrera, sheet 57–66, 1:200,000, Instituto Geologico y Minero de España (1970) and Jones (1984).

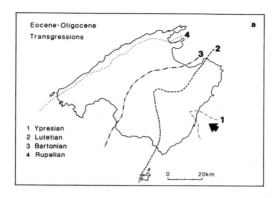

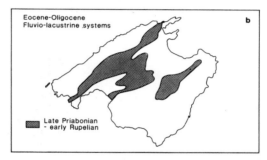

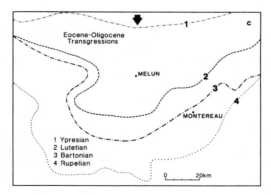

Fig. 16. Palaeogeographic schemes illustrating the hypothesized extent of various Eocene-Oligocene transgressions in Mallorca (a) and the distribution of fluvio-lacustrine deposits (b). Marine and freshwater sediments may be intimately associated. Note the similarity in the pattern of marine overstep between Mallorca and the Paris Basin (c). Data from Mégnien and Mégnien (1980) and Colom (1983).

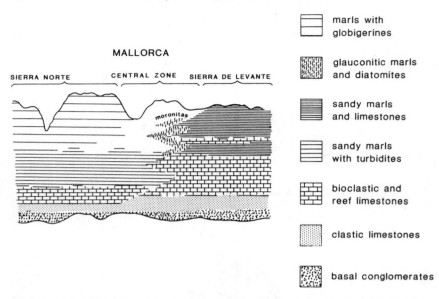

	marls with globigerines
	glauconitic marls and diatomites
	sandy marls and limestones
	sandy marls with turbidites
	bioclastic and reef limestones
	clastic limestones
	basal conglomerates

Fig. 17. Generalized scheme illustrating facies variation in the Burdigalian, after Colom (1975b). The basal conglomerate, not everywhere present, was deposited during a marine transgression. Overlying sediments include bioclastic and reef limestones (Sant Elm Calcarenites) and are also indicative of very shallow-water conditions. In the Sierra Norte these sediments are abruptly overlain by deep-water marls with turbidites (Banyalbufar Turbiditic Formation), implying rapid subsidence.

Tertiary sediments are mid-Eocene mammal-bearing lacustrine carbonates in the south-west part of the island and marine littoral calcarenites in the south-east. Lignite deposits, which have been commercially exploited, occur interbedded in the non-marine sediments along the south-east border of the Sierra Norte and the central part of the island (Pomar *et al.*, 1983b, 1985; Ramos *et al.*, 1985). These Eocene deposits have been interpreted as deposited in pull-apart basins located along strike-slip faults (Ramos *et al.*, 1988).

The marine units, rich in nummulites, were apparently deposited during a series of transgressions invading from the south-east and south-west. Colom (1983) recognised Ypresian, Lutetian and Priabonian (Bartonian) nummulitic deposits in the Eocene, and Rupelian (or Stampian) in the Oligocene (Fig. 16). The bulk of the Oligocene, however, is represented on Mallorca by continental conglomerates, breccias, sandstones, siltstones, lignites and limestones, locally containing large oncolites formed through the trapping and binding of sediment by filamentous cyanobacteria (Colom, 1975b, 1983): environments ranged from fluvial to lacustrine (Fig. 16). The pattern of marine transgressions and regressions, followed by non-marine environments, is remarkably similar to that established for the Anglo-Paris Basin (e.g. Gignoux, 1950; Mégnien and Mégnien, 1980), the Aquitaine and elsewhere and most probably reflects a dominant eustatic signal

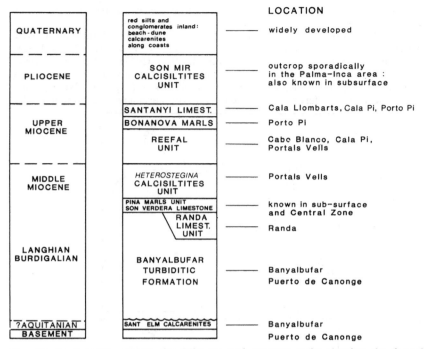

Fig. 18. Stratigraphic scheme for Miocene and younger rocks showing the formal classification erected by Pomar *et al.* (1983b) and Pomar *et al.* (1985), and the locations on the island.

although it could be attributed to coincidental parallel subsidence patterns of these areas.

From late Oligocene (Chattian) to early Miocene (Aquitanian) time there was deposition of continental clastics (conglomerates and sandstones, locally with oncolites, gastropods, ostracods) with local transient marine influence (Colom, 1976). Continental facies of this age also occur in Menorca. Lacustrine algal limestones of Aquitanian age are also known from western Mallorca.

A marine transgression of Aquitanian-Burdigalian age in Mallorca laid down the Sant Elm Calcarenites comprising reef and reef-associated facies that are laterally equivalent to fan-delta and lacustrine sediments. However, an ensuing sedimentary break heralded the onset of orogeny, and the Sant Elm Calcarenites are overlain in the Sierra Norte by bathyal clays and turbidites, locally containing resedimented Mesozoic blocks (e.g. Banyalbufar Turbiditic Formation: Rodríguez-Perea and Pomar, 1983). Hence the Burdigalian appears to have been a time of extensional faulting with local rapid subsidence; rhyolitic tuffs are recorded locally (Wadsworth and Adams, 1989). Lateral, more distal equivalents of the Banyalbufar Turbiditic Formation include the diatomaceous sediments and glauconitic marls of the Central Zone (Colom, 1975b). The facies variations and stratigraphy are illustrated in Figs. 17 and 18. Possible correlatives of these facies also occur on Menorca (Obrador and Fontboté, 1983).

In Randa (Central Zone, Mallorca) a thick sequence of proximal shelf-derived carbonates is dated as Langhian (Fig. 18; Pomar and Rodríguez-Perea, 1983). According to Mauffret *et al.* (1972) and Mauffret (1977) strike-slip faulting, initiated in the Oligocene, had brought Menorca from the north to the north-east of Mallorca by this time, displacing it some 75km (Fig. 19). Fossiliferous Palaeozoic pebbles, derived from Menorcan-type basement, are present in the

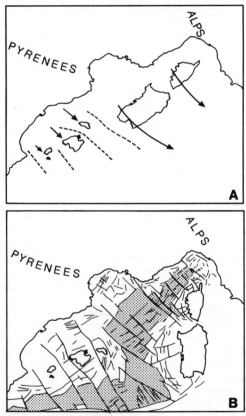

Fig. 19. A: Oligocene configuration of the western Mediterranean prior to the rotation of the Corsica-Sardinia block and the south-east translation of Menorca. Arrows indicate amplitude of future rotations; dashed lines major structural lineaments. Minor movement of Mallorca and Ibiza is also hypothesized to have taken place in the Oligocene. B: Present-day configuration of the Balearic area; dotted ornament indicates areas of stretched continental and/or oceanic crust. (After Mauffret, 1977; Durand-Delga, 1981; Rehault *et al.*, 1985). Note that such dramatic movement of Corsica and Sardinia is not accepted by all authors (cf. Biju-Duval *et al.*, 1978).

Burdigalian-Langian sediments of the Central Zone and the Sierra de Levante (Hollister, 1934).

In Langhian time the major orogenic phase took place, producing the Sierra Norte and the Sierra de Levante and the high-relief areas of Ibiza: Burdigalian sediments are themselves caught up in the deformation and the evaporitic substratum of the Keuper has commonly acted as a décollement surface. Subsequently, topographically lower ground was filled with molasse facies, including continental conglomerates, calcarenites and calcisiltites, and lacustrine

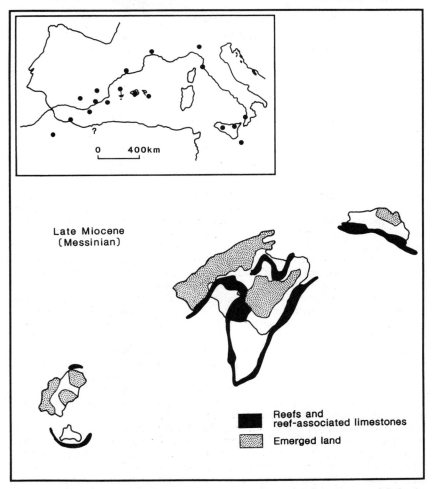

Fig. 20. Distribution of upper Miocene reefs in the western Mediterranean and Balearic Islands, after Esteban (1979/80).

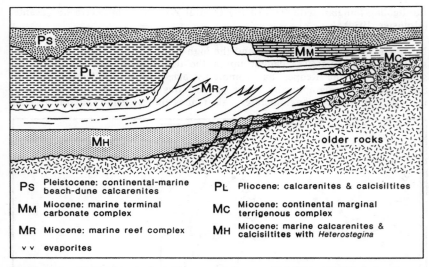

Ps Pleistocene: continental-marine beach-dune calcarenites

PL Pliocene: calcarenites & calcisiltites

MM Miocene: marine terminal carbonate complex

Mc Miocene: continental marginal terrigenous complex

MR Miocene: marine reef complex

MH Miocene: marine calcarenites & calcisiltites with *Heterostegina*

v v evaporites

Fig. 21. Facies distributions in the mid to upper Miocene of Mallorca, modified after Pomar *et al.* (1983c).

and alluvial-fan facies with mammalian fauna; these are dated as Serravallian-Tortonian and are little deformed (Pomar *et al.*, 1983b). These are broadly similar over all the Balearic Islands. Included amongst them are the Pina Marls and Son Verdera Limestone of Mallorca (Fig. 18), essentially lacustrine facies developed in the Central Zone and Sierra de Levante (Colom, 1967b; Pomar *et al.*, 1983b).

During the Messinian, evaporites accumulated in the offshore Balearic Basin (Hsü, 1977). Although not exposed on the islands themselves they are also recorded from wells in the Palma Basin. The uppermost Miocene is, however, spectacularly represented by reefs and associated shallow-water carbonates (Esteban, 1979/80; Pomar *et al.*, 1985). They are essentially undeformed and occur in belts around the islands; similar facies are known from elsewhere in the western Mediterranean (Fig. 20). In Mallorca, where these facies are best developed, the reefal unit typically overlies the Tortonian calcarenites (*Heterostegina* Calcisiltites Unit: Fig. 18) and includes shelf facies, reef-talus facies, reef-front and back-reef facies: all have high primary and mouldic porosity. There is also a marginal terrigenous complex ascribed to delta-fan, alluvial-fan, fluvial and lacustrine environments that crops out in the south-east of Mallorca (Fig. 21). True reefal facies are best displayed on the south coast of Mallorca (Cabo Blanco–Cala Pi: Fig. 22). Several prograding reefs may be recognized, implying changes in relative sea level (Pomar, 1988), and the older complexes are faunally more diverse than the younger. At Cala Pi coral genera forming patch reefs in a lagoonal facies include the forms *Tarbellastraea, Siderastraea* and *Porites*. At Cabo Blanco the corals at the base of the section, dominantly *Porites*, are dish-shaped, and evolve upwards to dish-and-stick and stick morphologies (Pomar *et al.*, 1985). Spur and groove tracts may be tentatively

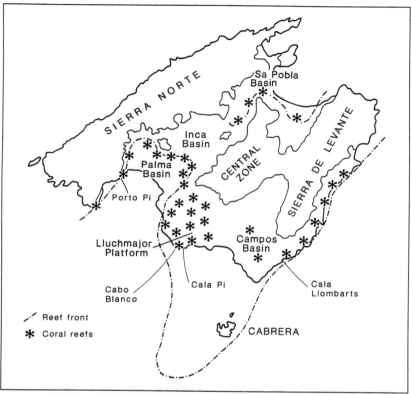

Fig. 22. Distribution of reef and associated facies in Mallorca, after Pomar *et al.* (1983c). The postulated extension of the reef front round Cabrera is based on geophysical evidence.

recognized. This decline in diversity could reflect increasing salinity in the Miocene Mediterranean as the draw-down began: unlike most corals, *Porites* today can tolerate salinities up to 45°/oo. Many of the karstic features on the island, including the development of caves, may have been initiated during this period.

Locally, above the reefal unit, lie the Bonanova Marls and conglomerates of fan-delta origin, succeeded by the Santanyi Limestones (Fig. 18), including oolitic, stromatolitic facies and calcilutites with rootlets interpreted as belonging to mangroves. The oolites, typically with herring-bone cross-bedding, are severely leached: they are best exposed near Cala Llombarts in Mallorca; the stromatolites, up to 5m in diameter and 2m high, are well seen at Porto Pi, near Palma de Mallorca (Pomar *et al.*, 1985: Fig. 22). Dated Pliocene includes grey marls and conglomerates: delta and bay environments are suggested (Son Mir Calcisiltites Unit: Fig. 18). Some beach-dune calcarenites are also assigned to the Upper Pliocene.

Marine cycle	Apparent sea level (in metres)	Faunal characteristics	Radiometric age
Z3	2	Banal	Post-Roman
Z2	2	Banal	
Z1	4	Banal	
Three aeolianite generations		HEMICYCLE B	
Y3	0.5–3	Probably banal	80,000 ± 5000 B.P.
Y2	1.5–2	Partial *Strombus* fauna	110,000 ± 5000 B.P.
Y1	9–15	Partial *Strombus* fauna	125,000 ± 10,000 B.P.
Two aeolianite generations		HEMICYCLE C	
X2	6.5–8.5	Impoverished Senegalese fauna	190,000 ± 10,000 B.P.
X1	2–4.5	Full *Strombus* fauna	210,000 ± 10,000 B.P.
Two aeolianite generations		HEMICYCLE D	
W4	4–8	Banal	> 250,000 B.P.
W3	15–18	*Patella ferruginea*	?
W2	22–24	*Patella ferruginea*	
W1	30–35	?	
Three aeolianite generations		HEMICYCLE E	
V(?2)	ca.15	Banal	
V(?1)	45–50	Banal	
Two aeolianite generations		HEMICYCLE F	
U	30(other levels?)		
?	60–65	*Patella ferruginea*	
?	75–80	(?*Purpura plessisi, Ostrea cucullata*)	
?	100-105		

Table 2. Sequence of glacio-eustatic Pleistocene cycles recognized from Mallorcan shoreline stratigraphy by Butzer (1975).

During the Pleistocene, the islands attained a form akin to their present-day configuration (Fig. 7). The temperature became much cooler than that of the Tertiary, and pluvial intervals heralded the glacial advances in the more northerly lands. Glacial maxima were marked by cryofracture and extreme aridity: some scree deposits descended to sea level (Butzer, 1964; Crabtree et al., 1978). As vegetative cover was destroyed, active fluvial down-cutting occurred, producing a broad spread of braided-river gravels over the central lowlands of Mallorca. Near the shores, there was deep incision of some valleys and widespread beach-dune development. Some of the coastal aeolianites are very steeply dipping: they contain abundant red algae and mollusc remains and are only feebly cemented. At least six glacio-eustatic littoral-sedimentary cycles are recognized, each cycle comprising a marine hemicycle with superimposed transgressive beach and interbeach capped by a clay-rich palaeosol, followed by a continental hemicycle consisting of several sets of colluvial silts each followed by an aeolianite and interrupted or followed by soil formation (Butzer, 1975). Three episodes of terra rossa pedogenesis on aeolianite parent material are known in Mallorca. A basic stratigraphy is shown in Table 2. During the mid-Pleistocene Riss glaciation (Fig. 7) sea level may have lain some 100m below its present level (Cuerda, 1975). During the interglacial interludes raised beaches accumulated and notches were cut in the cliffs (Fig. 8). Lower Pleistocene deposits (Calabrian) contain some cold-temperature bivalves such as *Arctica islandica*; and the warm-water gastropod *Strombus bubonius*, presently confined to the west coast of Africa, from Dakar southwards, is common in upper Pleistocene, particularly Tyrrhenian deposits of south-west Mallorca. In this area littoral environments were protected from north winds by the Sierra Norte and marine temperatures were relatively elevated (Cuerda, 1975; Pomar and Cuerda, 1979). The *Strombus* fauna first appeared no later than 220,000 years BP and was eliminated shortly after 100,000 years BP during a cool minor regression (Butzer, 1975).

As a consequence of the large amount of carbonate present in the ground waters, many of the Pleistocene and Holocene deposits on valley sides and in valley fillings in all of the islands are strongly cemented. At present, due to the strong seasonality of rainfall and the high summer temperatures, calcretes are still developing in many of the soil profiles.

Many of the limestone formations are penetrated by extensive cave systems. Scenically, some of the most celebrated caves in Europe are to be found close to the south-east coast of Mallorca. The Cuevas de Artá were cut in Jurassic limestones, whereas the more celebrated caves of Hams and Drach were formed in reef carbonates of the Upper Miocene. Many of the other extremely numerous

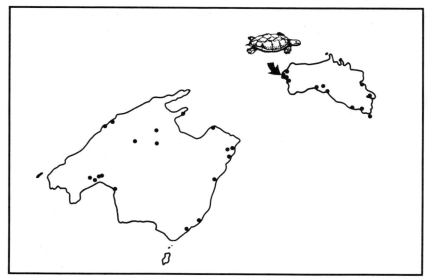

Fig. 23. Distribution of major terrestrial Pleistocene vertebrate remains in Mallorca and Menorca. Vertebrates chiefly comprise antelopes (*Myotragus*), shrews (*Nesiotites*) and dormice (*Hypnomys*). Note the occurrence of the giant tortoise *Testudo gymnesicus* in Menorca. Data from Cuerda (1975).

cave systems on the island have yielded Pleistocene mammal faunas, including abundant remains of antelopes, shrews and dormice (Bate, 1944; Cuerda, 1975); an unusual fauna of giant tortoises (up to 1 metre long) also developed during this time in Menorca and is known only from a restricted area in the west of that island (Fig. 23) (Bate, 1914a, b). The cavernous limestones of Mallorca also supply the island with abundant fresh water, even during the dry summer months. However, in the resort areas around Palma the tap water is becoming salty as the fresh-water lens is depleted by excessive use.

THE EXCURSIONS

EXCURSION 1
TRIASSIC FACIES AND TECTONIC STYLE

Locality 1: Coll de Sa Gremola, just north of km 106 post

From the Palma area take the Andraitx road, then follow signs towards Valldemossa and Soller, until the Coll is reached. There is space for cars to pull off the road. This exposure, tilted towards the north-east, cuts through red, brown and green marls containing much laminated gypsum typically white to grey but locally coloured black. A sample of black gypsum analysed for organic matter gave only 0.18% total organic carbon: it is not, therefore, bituminous. The dips are slight except in some zones of extreme deformation, where the evaporites are mylonitized: such fabrics are best seen on the rock face behind the west wall of the road cut. In the lower levels of the outcrop, probably close to the contact with the Muschelkalk, dolomitic beds occur. The laminated gypsum is best interpreted as a subaqueous deposit formed perhaps in a hypersaline lagoon or lake (Table 1). Spanish geologists generally assume a marine connection for Keuper facies in the Subbetic Zone (e.g. García-Hernández *et al.*, 1980), but evidence is lacking in Mallorca. The Keuper has commonly acted as a décollement horizon during emplacement of the nappes in the Sierra Norte which accounts for its localized extreme deformation. Its thickness is difficult to establish because of such tectonic complications but estimates as low as 50–80m have been made (Boutet *et al.*, 1982). Colom (1975b) suggested a maximum figure of 200m, and Rodríguez-Perea *et al.*, (1987) have measured up to 287 metres; these authors indicate that the evaporites themselves may locally attain 100m. The Mallorcan Keuper is faunally sterile apart from one possible reptile bone; in mainland Spain, however, it locally contains crustaceans of Late Triassic age.

Locality 2: Restaurante Es Grau

Opposite the restaurant car park, just above the small tunnel, a look-out point on Liassic platform carbonates offers good views to the north of the Triassic stratigraphy and tectonic style. The red Buntsandstein is visible at sea level in several bays: these sequences belong to the lowest unit in Fig. 5.

Locality 3: Estellenchs – Puerto de Estellenchs

A broad cultivated valley, containing numerous orange groves, stretches from Estellenchs down to the little harbour at the shore. It is perhaps easiest to leave vehicles in the village, just off the main road, and walk down, although a descent is possible with a car. At sea level and for some way above, particularly in a near-vertical cliff on the north-east side of the depression, an excellent exposure exists of red Buntsandstein and grey dolomitized Muschelkalk, although it is difficult of access. The Buntsandstein is best seen at sea level behind the retaining wall and on the south-west side of the valley where higher levels are typically discoloured.

Tabular and trough cross-bedded quartzose micaceous sandstones containing rip-up clasts and lignite (fragments of *Equisetum arenaceum*) are well displayed; lateral accretion surfaces may also be observed. Finer-grained (overbank) silty facies show interference wave ripples on bedding surfaces and are locally bioturbated. Certain clay-grade beds contain nodular calcrete, these latter being best seen along the coast-line; locally they are reworked. Blocks of bioturbated and rippled sandstone are also visible here. Two end-member facies associations may thus be distinguished: one fluvial-alluvial, represented by channellized sandstones and siltstones with calcretes; another lacustrine, represented by pervasively wave-rippled and bioturbated siltstones.

Note also the deposit of calcitic travertine partially capping a faulted block of Muschelkalk close to the south-west end of the bay. This calcitic precipitate locally includes large pebbles (10−20cm in diameter) and shows radial and tubular fabrics and botryoidal surfaces. Also of interest is the coarsely conglomeratic fill of a ?Pleistocene V-shaped stream valley that cuts through the red Buntsandstein behind the retaining wall. Blocks of Muschelkalk on the foreshore at the northern end of the bay contain molluscs, serpulids and carbonate pseudomorphs after gypsum. Sea-grasses (*Posidonia oceanica*), commonly encrusted by bryozoans, are visible in this and many other coves around the island.

Cross-bed directions of the Triassic channel sandstones, measured by Freeman (1975), suggest that the currents flowed south-east. Intriguingly, if Buntsandstein dispersal data are plotted for eastern Spain and Mallorca and Menorca, the sediment-transport directions fit uncannily well if the two islands are rotated back against the mainland (Fig. 24). Although tectonic rotation of nappes would render this interpretation invalid, it is compatible with a post-Triassic opening of the Valencia Trough (assumed to rift in the Chattian-Burdigalian interval), and near-concerted movement of a Mallorca-Menorca block. Although not directly dated in Mallorca a Werfenian age has been established for the Buntsandstein facies in mainland Spain (Busnardo, 1975). Dredging to the south of Mallorca has recovered red siliciclastics attributed to this unit, implying that the South Balearic Margin is continental, probably having rifted and foundered at the same time as the Valencia Trough (Curzi *et al.*, 1985). Buntsandstein facies, containing channel-lag conglomerates, are also well exposed at Puerto de Canonge (Locality 18). The recent work of Calafat (1986) and Calafat *et al.* (1986/87) has revealed the presence of reptilian vertebrates, fish, crustacea and insects locally in the Mallorcan Buntsandstein.

The maximum thickness of Buntsandstein is estimated as 250m by Hollister (1934) and Colom (1975b), 130m by Boutet *et al.* (1982), and 450m by Rodríguez-Perea *et al.* (1987); the base of the formation is not seen at Estellenchs, where it passes up gradually into the Muschelkalk, of which some 85m is exposed (Freeman, 1972). This has suffered virtually complete dolomitization, but a range of fabrics, including stromatolitic and fenestral, indicates peritidal deposition for the lower levels. Other horizons are abundantly bioturbated (presence of *Planolites*) and indeed elsewhere on Mallorca the Muschelkalk has yielded an abundant fauna of crinoids, calcareous algae, occasional gastropods, locally common bivalves and, rarely, and in more marly lithologies, the ammonoid *Ceratites* attributed to the Ladinian (Colom, 1975b: Table 1). The Ladinian, in

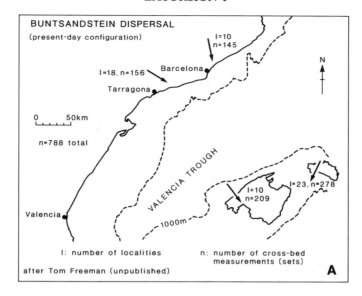

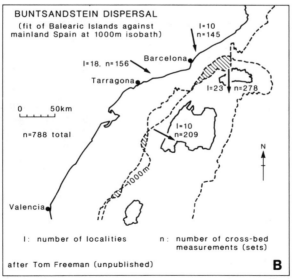

Fig. 24. Sketches showing palaeocurrent data for the Buntsandstein, after Freeman (1975) and his unpublished data (A and B). Note how the palaeocurrent directions for the islands align with those on the mainland after they have been rotated back against Spain by closing the Valencia Trough (B).

fact, seems to have been characterized by a relatively more deep-water Muschelkalk facies as it also locally contains abundant remains of the thin-shelled bivalve *Daonella*, common in pelagic limestones of the Alpine Trias and thought to have been epi- or pseudo-planktonic. The range of environments reconstructed for the Muschelkalk is illustrated in Fig. 25.

The dolomitization of the Muschelkalk bears no relation to sedimentary facies but is clearly joint-controlled. There is negligible evidence for de-dolomitization. Freeman (1972) suggested a late-stage process related to passage of meteoric fluids that had first passed through the Keuper and picked up elevated quantities of magnesium from clay minerals. However, it is difficult to see how such fluids would be sulphate-poor, an apparent prerequisite for dolomitization (Baker and Kastner, 1981; cf. Morrow and Ricketts, 1986) and this model may need re-examination. The dolomite problem, however, still remains largely unresolved as indicated by the recent review of Hardie (1987) whose favoured model is in line with that of Freeman (1972).

Recommended stop for sustenance: Hostal Maristel, Estellenchs.

Locality 4: Roadside north of Estellenchs

To the north of Estellenchs, just north of the km 93 post, good exposures of steeply dipping conglomerates of Burdigalian age may be seen. Their deformation attests to post-Burdigalian tectonism in the Sierra Norte. Evidence for early Langhian movement is presented at Locality 5.

Locality 5: Urbanizacion de George Sand, near Valldemossa

Some 4km to the west-south-west of Valldemossa, between the km 75 and km 74 marker posts, an access road leads west off the C710 to a new housing development: the Urbanizacion de George Sand. On the south side of the access road a cut rock face, excavated back from the road-side, elegantly illustrates the structural style prevalent in the Sierra Norte. Close to ground level are grey-green marls containing granular beds of coarser material; they contain a pelagic globigerinid microfauna and are dated as basal Langhian. The highest level of this Miocene deposit contains a 20cm-thick breccia of reworked Triassic material. Above this follows some 15 metres of dolomites and red, green and black marls. From one of the latter Boutet *et al.* (1982) have obtained a palynological assemblage of Norian age. Thus in this exposure dolomites attributed to the supra-Keuper (Table 1) are thrust over Miocene, and the inclusion of Triassic breccia in the uppermost levels of the deep-water marls implies that thrusting took place during early Langhian time. The Triassic shows only modest deformation with development of imbricate structure: the tectonic contact itself is nearly horizontal.

Locality 6: Roadside between Deià and Soller

Between Deià and Soller on the C710, close to the km 57 post, a small road leads seawards where a hoarding advertises the Restaurante Bens d'Avall, opposite which mylonitized gypsiferous Keuper is exposed next to a stone wall built of the evaporite. Gypsum and anhydrite have been recognized here. Some 50m down the road, which swings to the left, steeply dipping Aquitanian-Burdigalian, probably

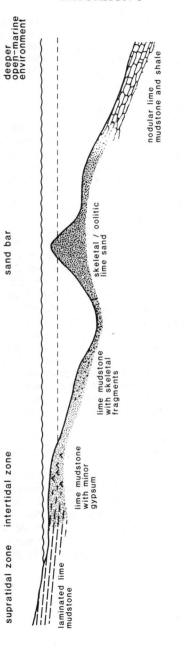

Fig. 25. Postulated environments of deposition of the Muschelkalk, modified after Freeman (1972).

corresponding to the Sant Elm Calcarenites, is again exposed behind a wire fence. Molluscs and shell fragments are common in this facies, but the roadside exposure is complex, with many included boulders of Muschelkalk or other carbonates which may represent a Pleistocene 'cold-climate' cemented hillwash. It should be noted, however, that Miocene deposits with exotic blocks have been described from this region by Pomar and Colom (1977).

Locality 7 (optional): Sa Calobra/Torrent de Pareis

Continuing along the C710, skirting Puig Major, the highest mountain on the island, one may descend seawards on a small road towards Sa Calobra along a spectacular series of hair-pin bends. Several hundred metres of platform carbonates are exposed here, extending, albeit with structural discontinuity, close to the summit of Puig Major. Tectonic repetition is indicated by the presence of Triassic red beds at several levels. Four thrust sheets were recognized by Rodríguez-Perea and Fornós (1986), who noted that the platform carbonates locally contain problematic breccias on various scales, interpreted as due to evaporite solution during Early Tertiary uplift and more recent karstic processes (Fornós et al., 1986/87; Adams, 1989). Centimetre-scale carbonate-filled vugs, probably replacing evaporite nodules, are also common at some horizons.

The typical karst topography in the Liassic platform carbonates is well displayed along the road, particularly between the km 7 and km 10 posts. Although the karstic limestone surface appears bare it actually supports a flora of algae, fungi and lichen which help corrode the rock; rates of surface dissolution are estimated as $1-2\text{cm}/10^3$ years (Pomar et al., 1975; Crabtree et al., 1978). Close to sea level the road divides and, on the left hand side of the road above the fork, highly weathered, brecciated igneous rock outcrops, probably a flow intercalated within the Keuper. (There is space for a car to pull in here). Some of the basalt is reddened and amygdaloidal, and locally contains jaspery material. Description of Triassic igneous rocks in Mallorca suggests that they are essentially basaltic, albeit extremely altered, with original mineralogy difficult to decipher: olivine, pyroxene and feldspars have, however, been recognized (Colom, 1975b). Dykes, sill, flows and pyroclastics have all been described (Rodríguez-Perea et al., 1987). Classically such rocks are held to testify to early rifting and graben formation in the western Mediterranean (e.g. Bosellini and Hsü, 1973; Bernoulli and Jenkyns, 1974).

Leaving vehicles at the car park next to the sea, the Torrent de Pareis is approached through a narrow, poorly illuminated tunnel: breccias are abundant in the platform carbonates at this level. This impressive canyon appears originally to have been a cave that was incised by a Pleistocene river when cutting the narrow gorge that today winds inland for several kilometres. At maximum the walls of this feature attain 300m in height and, locally, its width narrows to 5 metres (Colom, 1982). The storm beach at the mouth of the gorge is notable.

EXCURSION 2
DROWNED CARBONATE PLATFORMS, REDEPOSITED OOLITES AND JURASSIC PELAGIC FACIES

Locality 8: Recó and Cutri, Sierra de Levante

This outcrop, the most spectacularly exposed Jurassic section on the island, is approached from the newly widened road (C715) that runs between Artá and Capdepera, in the north-east of the island. Close to where a series of pointed electricity pylons crosses the road a track leads north along a valley: its entrance lies between the former positions of the 72 and 73km posts on the highway. (They were removed when the road was widened). Vehicles can usually be left at the roadside. An 18-hole golf course is now (1989) being constructed here, and permission may in future be needed to reach the outcrop. However, the regularly bedded sequence of platform carbonates that build the flank of Recó should be easily reached (Fig. 26).

The easterly (30°) dipping platform carbonates of Recó, built primarily of dolomites at the base and limestones at the top, display a range of peritidal fabrics: edgewise breccias, skeletal sands, pelletal sparites, gastropod-rich micrites, birdseye vugs and dolomitic white-weathering algal-laminated facies (Fig. 27). These latter occur rather regularly, typically in beds some 50cm thick, although they may be much thinner. Corals and hydrozoans also occur, being

Fig. 26. View of Recó and Cutri from the main road, indicating the suggested traverse. The view will change somewhat with the development of the golf course.

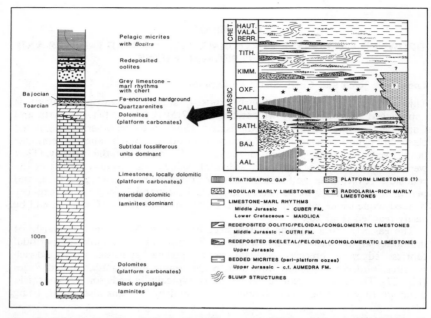

Fig. 27. Columnar section of Cutri, after Alvaro *et al.* (1984) and Prescott (1989), related
to the latter's postulated Mesozoic stratigraphy for the Sierra de Levante.

more common higher in the section. The vertical sequence of facies, representing
shoaling sequences of metre scale, reflects the balance between sediment
production, subsidence and changes in relative sea level (cf. James, 1984). Similar
facies of comparable age occur elsewhere in the Tethyan region (e.g. Pialli, 1971;
Colacicchi *et al.*, 1975). Although the shallowing effect is thought to be related
to sediment outpacing subsidence, the mechanism that initiates the new cycle,
with its subtidal facies, is not entirely clear.

From a vantage point near the top of this hill the higher sequence is beautifully
displayed on the western side of Cutri, displaced only by a few simple faults (Figs.
27, 28). The regularly bedded platform carbonates are overlain by a thick
sequence (ca. 50m) of more thinly bedded pelagic carbonates above which follows
a distinct mass of redeposited channelized oolite that varies in geometry along the
strike: this forms a massive cliff (Fig. 28). The thickness of the stratigraphically
continuous oolitic body locally ranges up to 30m. The oolite beds vary from
amalgamated sets to discrete channellized bodies and are interpreted as the
product of deposition in the channels of a submarine fan (Alvaro *et al.*, 1984).
Alternatively they may represent oolitic sand lobes deposited at the distal ends of
channels.

Descending into the stream valley and climbing up the other side the upper
levels of the platform carbonates are almost continuously exposed. Note the local
occurrence of quartz and calcite/saddle dolomite-filled vugs that represent

Fig. 28. View of Cutri, showing prominent ridge of redeposited oolite offset by an obvious fault. The lower regularly bedded unit is platform carbonates, overlying which, highlighted by snow, are thinner-bedded red and grey pelagic limestones of the Toarcian-Bajocian, in turn capped by the oolite of presumed Bathonian age. The lower more thinly bedded sediments directly beneath the ridge are constituted by interbedded oolitic and pelagic-bivalve-rich layers.

replacement of evaporite nodules and are suggestive of precipitation temperatures of 60–150°C (Radke and Mathis, 1980; Machel, 1987); anhydrite ghosts occur locally (Prescott, 1989). Close to the top of the sequence lenses of quartz granules and quartz sand become common. This siliciclastic facies is a marker horizon for the top of the platform carbonates and occurs also in the Sierra Norte. An orange-weathering dolomitic facies, some 40 metres thick, occurs at the top of the platform, but its contact with the overlying pelagic facies is poorly exposed; it may originally have been a carbonate sand, representing a relatively 'open' platform environment. Above, some 50cm of red pelagic micrite are visible below a 20cm-thick black, brown, yellow and red mineralized limestone. This latter facies contains numerous hematite-encrusted surfaces and pebbles and is rich in partly corroded ammonites: it is a typical Tethyan condensed sequence, classically interpreted as deposited on a topographic high or seamount (Jenkyns, 1971a). The following ammonites have been collected by us from this horizon:

FAUNA	SUGGESTED RANGE
Pleydellia mactra?	*Levesquei* Zone
Polyplectites discoides	*Thouarsense* Zone
Hammatoceras insigne	*Insigne* Zone

Hammatoceras sp.	*Bifrons* Zone and above
?*Grammoceras* sp.	*Thouarsense* Zone
Catacoeloceras puteolus	*Bifrons* Zone
Harpoceras exaratum	*Falciferum* Zone
Harpoceras sp.	*Falciferum* Zone or
	bifrons Zone
Partschiceras sp.	Long-ranging
Pleurolytoceras hircinum	Late Toarcian

This 20cm horizon thus represents the bulk of the Toarcian, reinforcing its interpretation as an extremely condensed sequence. (The Toarcian in North Yorkshire is about 125 metres thick; Cope *et al.*, 1980). It furthermore dates the drowning of the carbonate platform in this area as earliest Toarcian or latest Pliensbachian. The wider significance of this type of facies change from platform carbonate to pelagic, in terms of the relative roles of sediment accumulation, tectonism and eustasy, is discussed by Schlager (1981).

Above follows 1.8 metres of red nodular limestone (*ammonitico rosso*) of probable Aalenian or early Bajocian age. Above this, the grey pelagic limestones of the Cuber Formation, containing interbedded marls and replacement chert nodules, continue the sequence. Lower Bajocian ammonites occur (*Stephanoceras nodusum: sauzei* Zone: *Phaulostephanus* sp: *humphriesianum* Zone) at the base

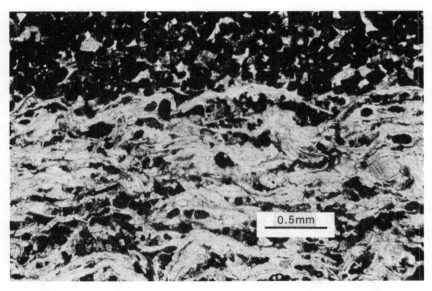

Fig. 29. Current-oriented shells of the bivalve *Bositra buchi* interlaminated with pelletal facies: such hydraulic sorting is common in redeposited pelagic sediments of mid-Jurassic age in the Tethyan region. Sampled from the base of the redeposited oolite, Bathonian, Cutri.

of this sequence and Bourgois *et al.* (1970) record upper Bajocian forms from higher levels. The supposedly pelagic bivalve *Bositra* is common in these limestones: a discussion of its lifestyle is given by Jefferies and Minton (1965) and Kauffman (1981).

Forming the bluff at the top of the hill is the mass of oolitic limestone, cut by a series of joints that are deceptively similar to high-angle cross-bedding. In basal, more thin-bedded levels, units of current-laminated *Bositra* shells and small pellets are interbedded with oolite: they occur in hydrodynamically separated layers (Fig. 29). The main mass of oolite shows a generalized grading and palaeocurrent data indicate a source area somewhere to the east (Prescott, 1989). It is attributed to the Bathonian. The ooliths themselves are both radial and concentric in structure, commonly show interpenetration, are accompanied by peloids and enclosed in interstitial calcite and dolomite of variable grain size (Fig. 30): diagenesis was presumably entirely submarine. The origin of the dolomite is unclear, but the magnesium may derive from expulsion of this element from originally high-Mg calcite ooids: the radial structure is suggestive of this primary mineralogy (Sandberg, 1975, 1983; Wilkinson *et al.*, 1985).

The occurrence of these redeposited oolites has until recently posed severe palaeogeographic problems as no source was known on Mallorca. Similar facies are recorded from several other localites in the Sierra de Levante and on the island of Cabrera (also Bathonian), the Central Zone (Randa: Anglada *et al.*, 1986) but

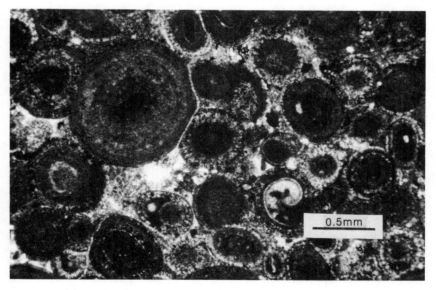

Fig. 30. Typical fabric of the redeposited oolitic facies. Radial and concentric ooliths are visible and interpenetration is common, suggesting appreciable pressure solution. Matrix is fine-grained calcite and dolomite. Bathonian, Cutri.

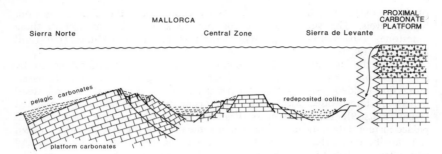

Fig. 31. Tentative palaeogeographic model for the middle Jurassic of Mallorca illustrating a suggested source for the redeposited oolites. The apparent absence of such derived shallow-water material in the mid-Jurassic of the Sierra Norte further controls the model adopted, by necessitating the presence of a submarine barrier in or close to the Central Zone.

they are absent from the Sierra Norte (Arbona *et al.*, 1984/5; Fornós *et al.*, 1984). Oolitic sediments of the same age also occur in the Prebetic and parts of the Subbetic of mainland Spain (Hoppe, 1968; Kuhry *et al.*, 1976; García-Hernández *et al.*, 1979, 1980). The mid-Jurassic oolites described from the Subbetic Zone, with which Mallorca is usually correlated, may themselves be redeposited; recent studies, however, suggest that some at least were formed more or less *in situ* (Molina *et al.*, 1984, 1985) although this may yet prove a contentious interpretation. The work of Bosellini *et al.* (1981) shows how large masses of mid-Jurassic oolite can be redeposited into pelagic basins with negligible evidence to betray their derived origin, apart from the presence of deep-water sediments stratigraphically above and below. In any event the recent recognition by Fornós *et al.* (1988) of some 60 metres of mid-Upper Jurassic oolitic sediments around the Artá Caves indicates a possible source region in Mallorca. Even if these oolites are ultimately recognized as redeposited, the Artá region must have been situated proximal to a carbonate platform. A model for the suggested palaeogeography is given in Fig. 31.

The presence of saddle dolomite in the top of the platform carbonates, indicating temperatures probably in excess of 100°C, coupled with the degree of interpenetration of the redeposited ooids, are suggestive of post-Jurassic burial depths of several kilometres in this sector of the Sierra de Levante.

Locality 9: Camp de Mar

Returning via Palma, Camp de Mar is situated west of the city, just south of the C719 on the road to Andraitx. Leaving the vehicle in the car park at the front, walk along the path that leads westwards just above the beach. Ultimately, the path peters out and grey limestone-marl rhythms can be examined at beach level. The occasional belemnite occurs in these facies. *Zoophycos* is common, as is *Chondrites* with darker piped into lighter sediment. Ammonites collected from these exposures include:

FAUNA	SUGGESTED RANGE
Haplopleuroceras	
subspinatum	*Discites* Zone
Dorsetensia sp?	*Humphriesianum* Zone
Graphoceras sp.	Upper Aalenian – Early Bajocian
?*Tmetoceras* sp.	Aalenian – Early Bajocian
Reynesella sp.	Early Bajocian
Oppelids	

The fossils indicate an early Bajocian age. These grey limestone-marl rhythms thus belong to the Cuber Formation (Fig. 12). Continuing westward, the path appears again where it crosses a fault and rises up a small flight of steps. The facies exposed here are pink in colour and variously constituted by crinoidal calcarenites (Fig. 32), stratigraphically overlain by brachiopod- and ammonite-rich coquinas and micrites. Spar-filled cavities, containing radiaxial fibrous calcite, occur locally: these cements are of probable submarine origin and are common in ancient reefs and mudmounds, as well as pelagic limestones that have undergone early lithification (Kendall, 1985). Much of the micrite is peloidal or pelletal and may in part represent a submarine cement (cf. von Rad, 1974, for a possible Recent analogue); the peloids were probably originally precipitated as

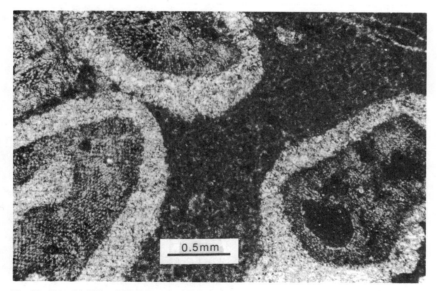

Fig. 32. Crinoidal biomicrite, showing well-developed syntaxial overgrowths: Pliensbachian-Toarcian, Camp de Mar. The pelletal structure of the enclosing micrite is reminiscent of certain high-magnesian calcite cements found on seamounts (von Rad, 1974). This crinoidal facies is typical for the open-marine facies on drowned but little subsided blocks of platform carbonates, i.e. pelagic seamounts (Jenkyns, 1971b).

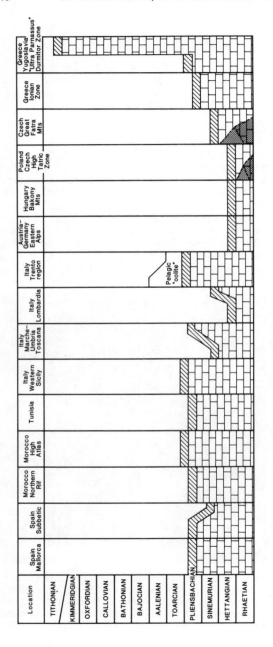

Fig. 33. Timing of facies change from platform-carbonate to pelagic around the Tethyan margins, after Bernoulli and Jenkyns (1974). These major, albeit diachronous 'drowning' events betoken changes in tectonic and/or eustatic patterns and are generally held to reflect rifting and foundering that predated sea-floor spreading in the Tethys-Atlantic System. Dating of the facies change is inexact and may differ within any one area in a complex way: the diagram is thus simplified. In the Subbetic Zone, for example, drowning events apparently took place locally as early as the Hettangian (García-Hernández et al., 1978).

high-magnesium calcite (Macintrye, 1985). The ammonites include harpoceratids, dactylioceratids and a *Hildoceras* suggestive of an early Toarcian age.

These open-marine facies stratigraphically overlie grey dolomites of carbonate-platform facies that, similarly to the Sierra de Levante, contain abundant quartz grains and granules. Hence, the 'drowning facies' of part of the Sierra Norte carbonate platform is exposed here: such sediments, commonly crinoid-brachiopod rich, are typical for all the Tethyan region (Jenkyns, 1971b). Recent parallels exist in and around certain shallow seamounts (Farrow and Durant, 1985), where crinoid-brachiopod-rich sediments accumulate at depths between 80 and 180 metres below sea level. In another area of Mallorca, close to Soller, and on the island of Cabrera, ammonite-bearing crinoidal facies of Pliensbachian age are known, possibly implying an earlier phase of drowning of part of the carbonate platform (Colom, 1973; Arbona *et al.*, 1984/5). Pliensbachian crinoidal sands are also common in the Subbetic Zone of mainland Spain (Dabrio and Polo, 1985).

Exact dating of the drowning event is not possible at Camp de Mar: it may be Toarcian but could be earlier. Thus whether there was a synchronous Pliens-bachian drowning event in all parts of Mallorca and Cabrera, with a sedimentary record preserved only locally or whether some blocks foundered in the Toarcian, is not established. This particular Mallorcan deposit recalls particularly the Brocatello of the Southern Alps in Switzerland (Wiedenmayer, 1963).

Around the Tethyan margins the time of drowning of the carbonate platforms, though diachronous, shows a distinct focus around Pliensbachian-Toarcian time (Fig. 33), and probably reflects a major rifting event, heralding the opening of the Central Atlantic-Tethyan seaway (Bernoulli and Jenkyns, 1974). On Cabrera there is locally an angular unconformity between the platform carbonates and overlying pelagic sediments which demonstrates the importance of extensional faulting at this time (Sàbat and Santanach, 1984). Actual drifting of America from Africa probably took place in Bathonian-Callovian time and, in tectonic terms, was a less spectacular affair (Winterer and Hinz, 1984).

About a kilometre west of this spot, at Cala Blanca, Wiedmann (1962) has described a Lower Cretaceous ammonite fauna from white chalky marls. Albian black shales are also exposed here. However, exposure is not particularly good and Cretaceous facies are better seen elsewhere (e.g. Locality 12).

Note that Butzer (1975) recorded a Pleistocene 'Senegalese' fauna of *Strombus bubonius* and other warm-water species, sandwiched between two generations of aeolianite, at Camp de Mar. The fauna occurs at +2m and key parts of the section are on an offshore islet in the centre of the bay (Crabtree *et al.*, 1978). There is now a bar on this islet and a causeway allows easy access.

Locality 10: Cala Fornells

A few kilometres to the east of Camp de Mar, south of the C719, is the attractive resort of Cala Fornells: the turn-off from the main road is at the west end of Paguera. After driving past a small cove, where Tithonian *ammonitico rosso* is prominently exposed, the vehicles can be left between the Hotel Cala Fornells and the Hotel Coronado, where the road ends (2.3km from Paguera). After walking back to the cove, platform carbonates, with quartz granules and grains, and locally showing gypsum pseudomorphs, may be seen on the left at the roadside and in the rock wall above, as can the contact with the overlying crinoidal calcarenites: the sequence is exactly comparable to that at Camp de Mar, although the exposure is not as illuminating.

At the far side of the cove there is a peninsula on which rests a house designed as a ship: below, platform carbonates are faulted against a dipping sequence of red-purple nodular *ammonitico rosso* overlain by regularly bedded white *Maiolica*. The sequence presumably crosses the Jurassic-Cretaceous boundary. Traversing along the outcrop toward the stratigraphically lowest level of the *ammonitico rosso*, a cleft in the rock, close to a fault, gives (difficult) access to the adjacent bay. Here, at sea level, various fabrics in the *ammonitico rosso* can be well examined, as can the fault plane itself. Some of the red-purple limestones are redeposited and retextured. The *Maiolica* here contains well-preserved aptychi. From this side also higher cliff exposures of the *ammonitico rosso* show evidence of slumping. Some remarks on this and a related Mallorcan section are given by Eller (1981). Underneath the *ammonitico rosso*, regularly-bedded grey limestones, here attributed to the Cuber Formation of Bajocian-Bathonian age, are seen close to sea level.

Returning through the cleft, down-section, a prominent iron-stained surface containing numerous ammonites is visible; this level may have been noted in the adjacent bay. This is an obvious hardground, whose formation is usually attributed to a pause in sedimentation.

Ammonites collected in the *ammonitico rosso* at this locality indicate a mid to Late Jurassic age:

FAUNA	SUGGESTED RANGE	
perisphinctid	?Late Jurassic	2 metres above iron-stained
Euaspidoceras sp.	Mid-Oxfordian	hardground
Cadomites daubenyi	*parkinsoni* Zone, latest Bajocian to *zigzag* Zone, early Bathonian	2 metres below iron-stained hardground

Thin-section study of samples of the *ammonitico rosso* facies reveal the presence of the planktonic crinoid *Saccocoma* in the higher parts of the red-limestone section, suggestive of a probable Kimmeridgian to Tithonian age (e.g. Turner, 1965; Mišík, 1966; Dromart and Atrops, 1988); stratigraphically lower samples contain abundant globigerinid foraminifera (Fig. 34), which are particularly common in the Oxfordian (Colom and Rangheard, 1966) and, as indicated by ammonites, the *ammonitico rosso* must extend through this level down to the Bajocian-Bathonian. Thus the Aumedrà Formation (Fig. 12) is missing here, as it is elsewhere in parts of the Sierra Norte (Alvaro *et al.*, 1984). The abundance of thin-shelled bivalves (*Bositra buchi*) in the lower parts of this section is also suggestive of a mid-Jurassic age (Jefferies and Minton, 1965), but these disappear abruptly at the iron-stained hardground which may therefore indicate the Bathonian-Oxfordian contact, with the Callovian missing as it is

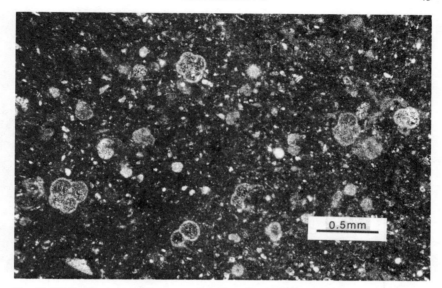

Fig. 34. ?Oxfordian *ammonitico rosso*, Cala Fornells. Abundant globigerinid foraminifera
are present: these disappear in the stratigraphically overlying *Maiolica* facies.

elsewhere in Mallorca (Fig. 12). That the higher levels of the *ammonitico rosso*
are of Tithonian age is also indicated by the local occurrence of the keyhole
brachiopod *Pygope* at this locality.

The *ammonitico rosso* is a facies particularly common in the Alpine-
Mediterranean domain, and widely used as a decorative marble: it floors
cathedrals as far apart as Mantua and Manila. Its estimated sedimentary rate is
in the range of a few mm/10^3 years and in such an environment nodular
submarine lithification has probably been important (Jenkyns, 1974). The source
of carbonate may have been aragonitic ammonites dissolving within the sediment
(Fig. 35). More distorted nodular structures have been produced when the
sediment has been transported and suffered soft-sediment deformation. If,
however, the nodules were well lithified when slumping and redeposition took
place, conglomeratic fabrics have resulted. Examples of this may be seen at
Comasema and Cuber (Localities 11 and 15).

The overlying white *Maiolica* limestone, nodular in its lower portion, continues
up into more regularly bedded limestones and marls. Unlike the *ammonitico
rosso*, which typically does not yield well-preserved calcareous nannofossils, the
Maiolica is formed substantially of the large coccolithophorid *Nannoconus*,
together with minor carbonate-replaced radiolarians and the tiny cup-shaped
ciliophores known as tintinnids or calpionellids: these latter are of considerable
stratigraphic use. *Maiolica* facies are known from the East and West Indies as well
as the Tethyan-Atlantic Ocean and its margins. In this section the lower levels of
these white limestones yield calpionellids and *Saccocoma*, suggestive of a
Tithonian age (cf. Turner, 1965).

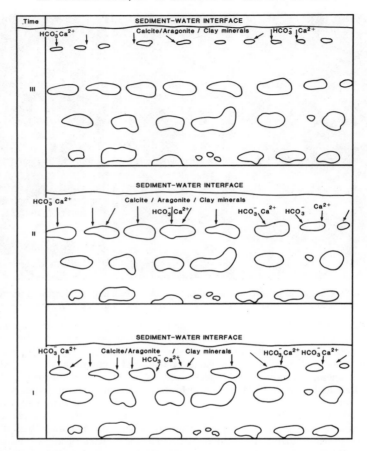

Fig. 35. Diagram to illustrate proposed rhythmic formation of nodules in *ammonitico rosso*. At Time I Ca^{2+} and HCO_3^- are being supplied by dissolution of aragonite and very fine-grained calcite near the top of the sediment column; the latest set of nodules is also close to the sediment-water interface, and hence the diffusion paths of the ions are short, and transfer rapid. At Time II the sediment-water interface has migrated upward, lengthening the distances between solution level and depositional sites, and thus slowing diffusion rates. At Time III a new series of nodules close to the sediment-water interface is being formed. Downward diffusion rates to the previous set of nodules became so low that Ca^{2+} and HCO_3^- were no longer removed as fast as they were supplied by the solution of aragonite and fine-grained calcite. An increase in the calcium carbonate concentration of the immediately sub-surface interstitial waters resulted, initiating a new set of nodules around skeletal calcite and micritic intraclasts. After Jenkyns (1974). This model is compatible with a primary difference in the carbonate content of the different layers, perhaps imposed by a climatic (Milankovitch-type) cycle (e.g. Fischer, 1986).

Of some interest is the complete lack of globigerinid foraminifera in *Maiolica* facies, even though they occur in beds stratigraphically above and below. Colom (1967a) ascribed this to dissolution at considerable depth, below the calcite compensation level for these particular forms, whereas the most resistant coccolithophores and calpionellids could survive. Other more palaeoecological interpretations are possible, but most authors suggest palaeodepths of several kilometres for this facies in keeping with Colom's interpretation.

A certain cyclicity is visible in the *Maiolica* facies and, indeed, in the *ammonitico rosso* itself, although to a lesser extent. Such a feature is common to most, if not all, pelagic sediments where they are uncontaminated by clastics (e.g. Arthur *et al.*, 1984; Barron *et al.*, 1985; Fischer, 1986). The cyclicity may manifest itself as a limestone-marl rhythm, an organic-rich – organic-poor couplet, a regular red-shale – green-shale alternation, and in many other ways, some of which may not be macroscopically visible. These cycles imply regular changes in: productivity of calcareous, siliceous and/or organic-walled plankton; dissolution of calcareous sediment; and influx of detrital clays, either singly or in combination. All of these parameters, directly or indirectly, may be related, via climatic change, to the orbital parameters of the earth: this is the so-called Milankovitch mechanism which controls variation in the intensity and latitudinal distribution of solar radiation through time. Major cycles of about 21,000, 40,000 and 100,000 years are predicted on theoretical grounds. All of these long-period cycles have been recognised in Mesozoic to Recent deep-sea sediments, although the most impressive record to date is from the Quaternary. The *Maiolica* cycles are probably best attributed to regular changes in plankton productivity, probably governed by current patterns and upwelling which in turn were climatically controlled.

Finally, after returning to the cars, note the Pleistocene dune calcarenites at beach level, opposite the Hotel Cala Fornells.

Having left Cala Fornells and passed the cove exposing the section just visited, the road swings left to follow a new one-way system. The new road, which climbs up a small hill, exposes a highly faulted Jurassic sequence of platform carbonates, a black ferromanganese-encrusted hardground, *ammonitico rosso* and the grey limestone-marl rhythms of the Cuber Formation. Note particularly the neptunian dykes and sills of pink crinoidal limestone (?Pliensbachian-Toarcian in age) that cut through the platform carbonates; they testify to syn-sedimentary tectonics and are common throughout the Tethyan region (Wendt, 1971). Intrusion of such crinoidal sediment into fractured platform carbonates apparently accompanied the regional break-up of many Early Jurassic carbonate platforms (Jenkyns, 1971b), hence fissures of Pliensbachian-Toarcian age are particularly common (cf. Fig. 33).

Recommended evening rendezvous: Bar Abaco, San Juan, 1, Palma de Mallorca: a rather special environment.

EXCURSION 3
PLATFORM CARBONATES, PELAGIC SEDIMENTS AND BLACK SHALES

Locality 11: Comasema (may be interchanged with Stop 15)

After following the main road from Palma to Soller (C711) for some 14km, take the turn to the right through Bunyola towards Orient. Cross over the pass and descend towards the valley: this is one of the more unspoiled parts of the island. Just before Orient, take the narrow road on the left towards Comasema. The road is gated, and care must be taken as this is private property. Once arrived at the large farm house at the end of the road, permission must be sought to continue (walk through the courtyard to the main door). Leave the cars at this point and, descending slightly and skirting the house, walk in the same general north-easterly direction on foot along the edge of the fruit-tree plantation. Note the near-verticality of the strata in the central part of the valley. Along the left side of the track, good examples of Tithonian *ammonitico rosso* in chaotic pebbly-mudstone facies, and containing included pelagic-limestone blocks up to 0.5m in diameter, ammonites, aptychi and belemnites, are well displayed. These deposits have been redeposited, probably as débris flows triggered by syn-sedimentary fault movements. Climb through the olive groves in a south-south-easterly direction towards the bluff illustrated in Fig. 36.

The geology of the bluff has been interpreted by Pomar (1976) as a series of slid blocks in a basinal pelagic sequence. A tectonic interpretation is possible and

Fig. 36. View of the bluff at Comasema, where a repetitive sequence of Jurassic facies is exposed.

perhaps preferable. With the knowledge gained on Excursions 1 and 2 it should be possible to recognize all the sedimentary units present and attempt an interpretation of the geology. Note particularly the well-displayed carbonate pseudomorphs after gypsum and quartz granules in the uppermost levels of the platform carbonates.

Recommended stop for sustenance: verandah of Hostal de Muntanya, Orient.

Locality 12: Road-cut near Orient

Some 2km east of Orient, between the km 13 and km 14 posts, the road descends through a shaded area with a cut rock face on the right-hand side. At road level white-weathering black shales are exposed: these are the Albian black shales (locally termed blue marls), rich in pyritized ammonites, pyrite nodules, and locally containing baryte and plant remains. Fish skeleta are common in this facies (Colom, 1984). Organic-carbon values are typically a few percent but one sample analysed from this outcrop gave 5.34% total organic carbon. The thickness of this unit is estimated as 25−30 metres in Mallorca (Alvaro-López *et al.*, 1982) but its base is not seen at this outcrop. Aptian-Albian black shales are particularly widespread, not only in the Alpine-Mediterranean region but in many parts of the globe. In the Subbetic of mainland Spain they are locally interbedded with oolitic turbidites (Kuhry, 1975). Carbon-isotope evidence demonstrates that this part of the Cretaceous saw anomalously high burial rates of organic carbon in the sedimentary record, and this has been referred to the influence of an Oceanic Anoxic Event (Schlanger and Jenkyns, 1976; Jenkyns, 1980; Scholle and Arthur, 1980). Such events were characterised by expanded and intensified oxygen minima impinging on many shelves and continental margins, conditions particularly favourable for deposition of carbon-rich sediments (Fig. 37).

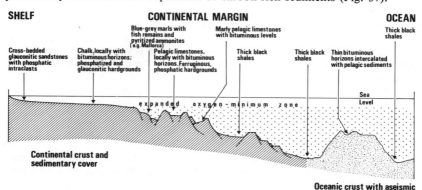

Fig. 37. Facies distribution across a continental margin during the Aptian-Albian Oceanic Anoxic Event: suggested palaeo-position of Mallorca indicated; after Jenkyns (1980). Mallorcan and Subbetic black shales are similar to those encountered in the Vocontian Trough, southern France, which also derives from the northern continental margin of Tethys. In the Vocontian Trough three particularly organic-carbon rich levels occur in the Aptian-Albian (Bréhéret, 1988).

Locality 13: Lloseta Quarry

This quarry, situated just west of the road between Lloseta and Biniamar, just south of Biniamar itself, shows a spectacular bedding surface of a well-developed Toarcian hardground, mineralized with yellow goethitic pisoliths and nodules rich in iron (40–50%) and titanium (0.3–0.6%) capped by a laminated black iron-manganese-titanium–rich crust (Alvaro et al., 1984; Prescott, 1988); elemental values of the crust typically range between 25–40% for Fe; 3–5% for manganese and 0.5–2% for titanium. Thin-section study reveals the presence of abundant encrusting foraminifera in this black mineralized layer. Ammonites collected by us from this hardground include:

FAUNA	SUGGESTED RANGE
Catacoeloceras confectum?	*Bifrons* Zone
Hammatoceras sp	*Bifrons* Zone and above
Harpoceras falciferum	*Falciferum* or *bifrons* Zone
?Dumortieria sp.	*Levesquei* Zone

This level probably embraces most of the Toarcian, and is thus comparable to that exposed in Cutri, in the Sierra de Levante. Similar mineralized facies, all comparable to present-day manganese nodules and crusts, are described by Jenkyns (1970, 1977), Germann (1971), Drittenbass (1979) and Mindszenty et al. (1986) from Jurassic pelagic limestone elsewhere in the Tethyan region. Iron-rich hardgrounds occur at several stratigraphic levels, including the Toarcian, in the Jurassic of the Subbetic Zone, mainland Spain (Barthel et al., 1966; Seyfried, 1978, 1980).

The mineralized facies in this locality are underlain by pink micritic pelagic limestones and locally dolomitic platform carbonates, the latter being well exposed at the western end of the quarry: quartz granules have not been found here. Neptunian dykes of pink biomicrite intrude the platform carbonates. The sequence is much disturbed by faulting but, above the hardground, ?Aalenian *ammonitico rosso* is exposed, showing well-preserved *Thalassinoides* and *Zoophycos*, overlain in turn by the Bajocian-Bathonian grey limestone-marl rhythms of the Cuber Formation. In 1987 this quarry was being filled by the dumping of rubbish and its future is hence uncertain.

Locality 14: Es Barraca

This section in platform carbonates is situated along the road that runs north from Inca, towards the north coast: the profile occurs above and below the km 11 post some 6 km north of Selva. However, the similar facies exposed on Recó, Sierra de Levante (Locality 8) display the sedimentary features of peritidal carbonates in a clearer manner. The Es Barraca section shows a series of shallowing-upwards cycles with development of oolitic, crinoidal, pellet-mud, and algally laminated facies. Supratidal facies are clearly indicated by the local presence of desiccation cracks, edgewise breccias and the pyramidal structures known as tepees (Alvaro et al., 1984). Dolomitized burrows occur at some levels, but the bulk of the formation is limestone. Almost opposite the km 11 post but

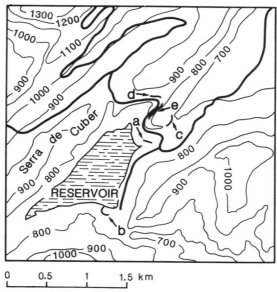

Fig. 38. Map of the Cuber area, showing location of key outcrops.

Fig. 39. Cross-bedded quartz-rich platform carbonates, Cuber. Such facies are characteristic of the uppermost levels of the platform sediments, below the break-up unconformity.

further along the road, there is a faulted sliver of the mineralized Toarcian hardground overlain by the grey limestone-marl rhythms of the Cuber Formation.

Recommended stop if returning to Palma at this point: Binisalem, where the José Ferrer Winery makes some of the best wine in the Balearic Islands. The vino tinto reserva, available in most shops, is worth sampling. (Also well worth seeking out, but more difficult to find, are the wines, largely made from French grape varieties, of Jaume Mesquida of Porreres).

Locality 15: Cuber

This stop is an alternative to Stop 11 (Comasema) and is rather more accessible. From Es Barraca continue northward until the C710 is reached and drive westward towards Soller. Park the vehicles close to the km 34 post, adjacent to a reservoir (Fig. 38) (a). You will notice a drainage trench running into the reservoir which exposes a near-vertical overturned sequence through the limestone-marl rhythms of the Cuber Formation (Alvaro et al., 1984). After examining this, walk round the side of the reservoir towards the dam: just before you reach it, on the left, there are excellent exposures of the cross-bedded quartz-

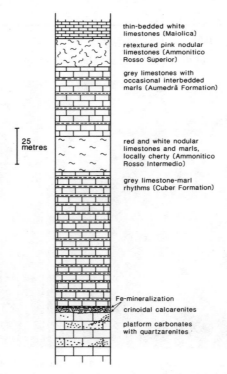

Fig. 40. Synthetic section of the Cuber area, modified after Alvaro et al. (1984).

arenites that characterize the top of the carbonate platform (Fig. 39). Proceed across dam into quarry (b). Note the overfold with inverted platform carbonates stratigraphically overlain by dark shell-rich limestones and black shales. These dark shales may be a reflection of the early Toarcian anoxic event described from various Tethyan localities (e.g. Jenkyns, 1988) but hitherto unrecorded from Spain (Prescott, 1989). This event left an organic-carbon-rich record in many Tethyan basins and seems to be related to the widespread development of intense oxygen minima impinging on large tracts of sea floor. It is thus similar in kind to the better documented Cretaceous anoxic events (Jenkyns, 1985).

Return to the vehicles. Walk up the road and then climb the bluff, parallel to the fence, proceeding in a northerly direction (c). The top of the bluff lies close to the core of a syncline formed by Tithonian *ammonitico rosso* and *Maiolica*. From this vantage point another reservoir lying some kilometers to the north-east is visible. Good exposures exist here of the *ammonitico rosso* in chaotic pebbly-mudstone facies containing many rolled and fragmentary ammonites: this facies is identical with that found at Comasema, some 4km (as the crow flies) to the south. These retextured sediments presumably attest to local syn-sedimentary tectonic activity at the close of the Jurassic.

Drive a kilometre up the road and park by the hair-pin bend just before the km 35 post. On the roadside (d) the top levels of the platform carbonates are again exposed and contain the diagnostic quartz sand and granules. Above, there is a yellow-stained hardground overlain by 2.5m of pink crinoidal calcarenites in turn capped by another ferruginous hardground. Walk to the other side of the hair-pin (e), crossing a fault. Here the *ammonitico rosso intermedio* (Bathonian/Oxfordian) is exposed at the roadside, locally chert-bearing and containing numerous *Bositra* shells. Much of it is discoloured. The Cuber Formation is developed stratigraphically below, along the road, and the Aumedrà Formation stratigraphically above in the hill on the left. A synthetic section of Jurassic stratigraphy in the Cuber region is given in Fig. 40.

EXCURSION 4
PALAEOGENE NON-MARINE LIMESTONES AND
LOWER MIOCENE REEFS AND TURBIDITES

Locality 16: Costa de la Calma

A few kilometres to the west of Palma Nova is the southwards turn-off to Costa de la Calma. The section is examined along the coast, a little to the north-west of the settlement, walking westwards towards Paguera (Figs. 41–44). Detailed descriptions were given by Colom *et al.* (1973), Marzo *et al.* (1983), Jones (1984), and Ramos *et al.* (1985), who interpreted the sequence as largely constituted by lacustrine, fluvial and alluvial-plain facies, an interpretation generally in accord with earlier views. Two main lithostratigraphic units can be differentiated in the Palaeogene of this area of Mallorca: a lower unit of Eocene age (Bartonian-Priabonian) disconformably overlain by an Oligocene (Rupelian-?Chattian) unit (Fig. 45). Both units are visible in this coastal section. Much limestone

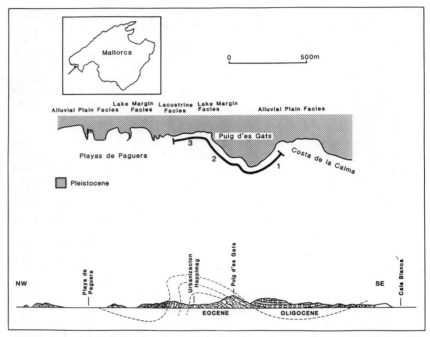

Fig. 41. Location of the Palaeogene non-marine facies close to Costa de la Calma, after Jones (1984). Structural cross-section after Ramos *et al.* (1985).

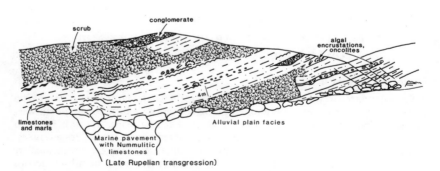

Fig. 42. Sketch of the cliff, west Costa de la Calma, after Jones (1984).

2 Puig d'es Gats

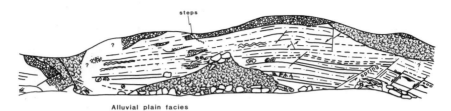

Fig. 43. Sketch of the cliff below Puig d'es Gats, after Jones (1984). Legend as in Fig. 42.

3 East Playas de Paguera

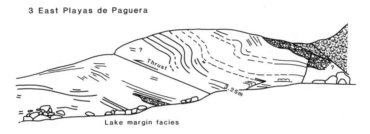

Fig. 44. Sketch of the eastern part of the Playas de Paguera, after Jones (1984). Legend as in Fig. 42.

conglomerate, full of light- and dark-coloured pebbles, chiefly constituted by Muschelkalk and platform carbonates, is visible at beach level and can locally be seen in stratigraphic contact with the sediments in the cliff section: these belong to the Oligocene unit. Note particularly the pressure-solution indentations in the pebbles where they have been in contact with each other.

The cliff section along the beach is dominantly calcareous with variable quantities of marl that contain charophytes (and locally mammal and crocodile remains and fresh-water gastropods) and a little secondary gypsum. Point-bar accretion deposits are visible in parts of the cliff. Lignite occurs at some levels in the sequence but undoubtedly the features of major interest, well seen on the west side of Costa de la Calma, are the levels of oncolites. These concentrically laminated bodies are of centimetre scale (Fig. 46) and are packed together in discrete beds that show some degree of grading: such algal-coated concretions are known from numerous Recent and ancient lacustrine and fluviatile environments (e.g. Walter, 1976). Note also the oncolitic coatings on twigs and plant remains. This part of the section is interpreted as an alluvial/delta-plain deposit by Jones (1984) and Ramos et al., (1985); the limestone conglomerates are thus assigned to channels. At the headland below Puig d'es Gats there is an obvious rock-fall of limestone conglomerate. At this spot, at sea level, there is a sandy marine

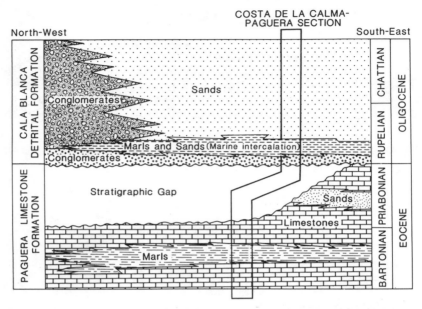

Fig. 45. Stratigraphy of the Costa de la Calma-Paguera area, after Ramos *et al.* (1985). Note that the clastic material is dominantly composed of limestone and dolomite.

Fig. 46. Oligocene non-marine oncolites; cliff below Puig d'es Gats. Diameter of coin = 2.5cm.

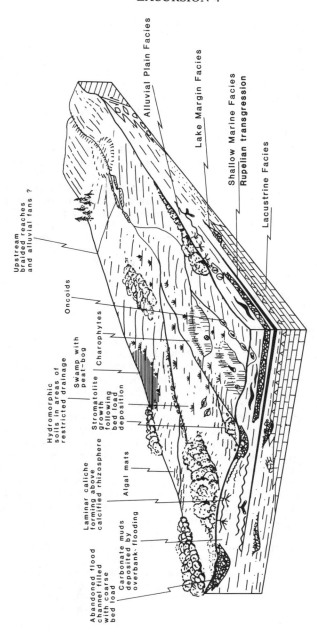

Alluvial Plain Facies

Lake Margin Facies

Shallow Marine Facies
Rupelian transgression

Lacustrine Facies

Upstream
braided reaches
and alluvial fans ?

Oncoids

Hydromorphic
soils in areas of
restricted drainage

Swamp with
peat–bog

Charophytes

Stromatolite
growth
following
bed load
deposition

Laminar caliche
forming above
calcified rhizosphere

Algal mats

Abandoned flood
channel filled
with coarse
bed load

Carbonate muds
deposited by
overbank flooding

Fig. 47. Depositional model for the Palaeogene facies of Costa de la Calma and Paguera, after Jones (1984).

intercalation of Rupelian (Stampian) age containing numerous nummulitids and much bivalve, gastropod, echinoid, crinoid and coral débris. The sediment is pervasively bioturbated. Marine elements also occur in some of the stratigraphically associated marls. This marine horizon was deposited during a major marine transgression which covered much of the island of Mallorca. As it contains derived boulders of fresh-water limestones it may perhaps be best interpreted as a littoral deposit. This transgression also affected the Anglo-Paris (Fig. 16) and Aquitaine Basins and most probably reflects a eustatic rise in sea level.

The section continues round the headland, although much disturbed by faulting, to a small bay. Note particularly the red and yellow-coloured clastic-carbonate facies, locally channellized, cross-bedded and containing many algally encrusted branches and twigs up to 60cm long and 20cm across. Calcareous concretions that resemble calcrete occur in some of these red clastic carbonates, and this part of the sequence has been interpreted as fluvial by Colom et al. (1973). Jones (1984), however, prefers an interpretation as a lake-margin facies, and disputes the presence of calcrete. Her generalized facies model is given in Fig. 47. The traverse can be stopped at this point, or can continue through similar facies, but belonging in part to the lower, Eocene unit and interpreted as lacustrine lake-margin and alluvial plain, to Paguera (Fig. 44).

Locality 17: Banyalbufar

From Costa de la Calma rejoin the C710 and follow the road via Andraitx, Coll de Sa Gremola, Estellenchs to Banyalbufar. Vehicles may sometimes be left in front of the Hotel Mar y Vent. Descend on foot along the road in front of the hotel, then turn hard right, passing behind it, and continue down through the terraces towards the little harbour, until the surfaced road finishes by a stream course and becomes a path through the pine woods. Do not descend to the jetty and beach at this point where a left fork is possible but walk northwards a little way along the higher path through the pine wood: the path becomes narrow and liable to subsidence. Where the path descends to beach level, examine the spur of rock that runs into the sea. It is largely constituted by black oil-stained Muschelkalk that gives off a petroliferous odour when hit with a hammer. The upper surface is penetrated by numerous *Lithophaga* borings and is covered with a veneer of a light-coloured carbonate that contains many angular blocks of Muschelkalk. Some borings contain fills of this light-coloured carbonate. This surface represents an unconformity between Triassic and Miocene and repays careful study. Note the red algae, clams and corals. Note also that the cliff exposure landward of this outcrop shows bedding discordances. Looking north-east a channellized turbidite body is also visible.

Retracing one's steps to a point above the jetty (the 'left' fork alluded to above), a descent is possible down a steep path, past a house. The geological relationships are tolerably clear at this point. At beach level there is black oil-impregnated Muschelkalk, rich in trace fossils, upon which, just north of the jetty, a sequence of blue marls and calcarenitic turbidites rests directly (Banyalbufar Turbiditic Formation). This sequence will have been encountered earlier in the descent. The basal 'sandy' facies, where it rests on the Muschelkalk, contains pebbles of a derived light-coloured calcarenite. Most prominent in the cliff is a redeposited channellized unit with a sharp base, apparently a composite bed at first glance, that is obviously graded with rip-up clasts and laminated and ripple-bedded at the top where it passes up into marls. This turbiditic mass thins from some 5 metres at the point of descent to some 2 metres at its northern extremity above a coral-red algal carbonate unit (Sant Elm Calcarenite Formation) that is intercalated between the Muschelkalk and the deep-water Miocene sequence (Fig. 48). It also thins to the south. On close inspection the composite layering in the channellized turbidite appears to be largely due to differential cementation, but this probably reflects original bedding, for the unit's internal geometry downlaps to the north-east. This bed contains derived shallow-water débris including large foraminifera: planktonic elements, namely micro- and nannoplankton, indicate a Burdigalian-Langhian age (Bizon *et al.*, 1973; Rodríguez-Perea and Pomar, 1983).

The reef facies (Sant Elm Calcarenite Formation), alluded to above, is well seen to the north of the jetty. Here it appears to be a solid framework built largely of encrusting red algae (Figs. 49, 50), many of which are still rose-coloured, oysters and scleractinian corals. Derived Muschelkalk clasts up to 1m in diameter are locally common in the reef rock and it is apparent that there was considerable palaeo-relief on the Muschelkalk surface prior to the deposition of the younger coralliferous carbonate. The contact between this sediment and the overlying Banyalbufar Turbiditic Formation is unconformable, and locally erosional. These Sant Elm Calcarenites, dated as Aquitanian-Burdigalian in age, were clearly laid

Fig. 48. Discontinuous body of reef carbonate (Sant Elm Calcarenite Formation) unconformably overlain by the Banyalbufar Turbidite Formation which is constituted by a series of marls and redeposited quartz-bearing calcarenites: lower Miocene. Note the pinch-out of the turbidite body against the reef rock. Bay below Banyalbufar.

Fig. 49. Red algae, Sant Elm Calcarenites: basal Miocene. Bay below Banyalbufar. Diameter of coin = 2.5cm.

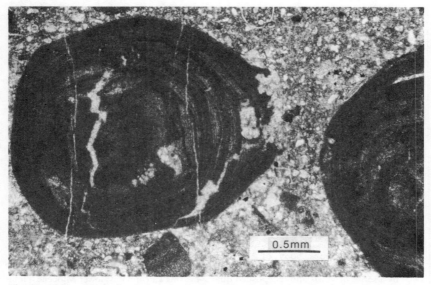

Fig. 50. Thin section of red algae in the Sant Elm Calcarenites. Bay below Banyalbufar.

down close to shore in very shallow water, yet the overlying turbiditic series accumulated at considerable depth. This implies very high local rates of subsidence during Burdigalian time, which is reflected in the wide variety of facies of this age on the island (Fig. 17). A horst-graben pattern of sedimentation was inferred by Pomar (1979), probably related to extension and crustal thinning in the Valencia Trough (Riba, 1983); alternatively the turbidites may be genetically related to a growing orogenic wedge (Ramos *et al.*, 1988).

A possible source rock for the oil that has migrated through the Muschelkalk would be the overlying Banyalbufar Turbiditic Formation. Analyses of shale samples from this series for organic carbon gave values between 0.3 and 0.9% total organic carbon (Fig. 51), which is suggestive in this respect. Partial

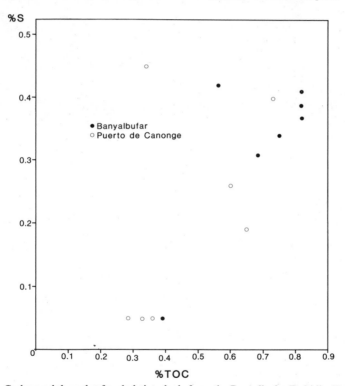

Fig. 51. Carbon-sulphur plot for shale interbeds from the Banyalbufar Turbidite Formation from Banyalbufar itself and Puerto de Canonge. This type of pattern, with a regression line that would pass close to the origin, is typical for sediments deposited under normal marine (*not* euxinic) conditions (Berner and Raiswell, 1983). The two samples that contain anomalously low contents of carbon have probably suffered surface weathering. Correlative sediments laid down on the northern side of the Valencia Trough were probably deposited under more oxygen-deficient conditions (Demaison and Bourgeois, 1985).

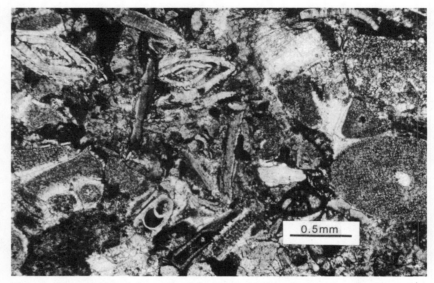

Fig. 52. Thin section of Sant Elm Calcarenites, showing the lime sand to be rich in echinoderms, foraminifera and mollusc shells. Cliff below Banyalbufar, south-west of the bay.

correlatives of this unit, but more organic-rich, are generally thought to have sourced the oil on the northern side of the Valencia Trough where hydrocarbons have been found in karstic cavities excavated out of Mesozoic carbonates overlain by Miocene sediments (Stoeckinger, 1976; Demaison and Bourgeois, 1985; Albaigés *et al.*, 1986). However, the Sant Elm Calcarenites are not apparently oil-stained, implying either that migration took place before their deposition or that the passage of oil has left no obvious trace. Mesozoic sources have been proposed for the oil in the Valencia Trough, and may also have to be considered for the staining of the Mallorcan sequences. Retracing one's steps southward, passing the jetty and boathouses, walk round the bay up the steps on the west side. The Muschelkalk is stratigraphically overlain by some 15cm of blue marl capped by a lens of conglomerate: i.e. the Sant Elm Calcarenites are also missing here and the Miocene turbiditic formation rests directly on Trias. Looking back at the cliff it is apparent that the large turbidite body here displays downlapping geometry towards the south-west. At the top of steps on the cliff top the calcarenites are well-developed, however, but lack reef structure. Occasional corals are present, as are abundant red algae, bivalves (oysters) and derived Muschelkalk clasts. Walking westwards the sediment becomes a more uniform lime sand (Fig. 52) with fewer macrofossils: echinoids and sand-grade Muschelkalk fragments are notable.

From this cliff-top vantage point the general geological relationships can be viewed and discussed. The major features are:

1) An oil-impregnated Muschelkalk surface with considerable (metre-scale) palaeorelief.
2) A Burdigalian hardground with associated reef development.
3) A Burdigalian peri-reefal calcarenite.
4) A Burdigalian to lower Langhian turbidite facies that locally rests on 1, 2 and 3 above.
5) Question: did a Burdigalian reef rim run across the present mouth of the bay? Was it close to the palaeo-shoreline?
6) Question: was a broad channel excavated in the Sant Elm Calcarenite Formation and later filled with the Banyalbufar Turbidite Formation?
7) Question: is the erosion surface between these two formations submarine or subaerial?

Walking back towards the path from this point the differentially cemented turbidite, showing numerous centimetre-scale pebbles above a scoured base, may again be examined.

Locality 18: Puerto de Canonge (optional)

From Banyalbufar continue eastwards along the C710 for a little over 7km to the km 80 post and then take the narrow road, with many hair-pin bends, that leads seaward and downward, following signs to Puerto de Canonge (Port d'es Canonge). Park at the base, in the settlement, and walk westwards. Red Buntsandstein is well-exposed in cliffs and spurs running out to sea. Continue walking parallel to coast along the footpath, crossing a small stream bed, until a wide track is reached, the seaward side of which is built up with large stones. Proceed along this track, noting the coarse conglomeratic lenses in the Buntsandstein and the levels of local discolouration. The contact with the Sant Elm Calcarenites is tolerably well exposed, although the section is locally inverted. A prominent peninsula of near-vertical Miocene sediments juts into the sea beyond this point; as the path swings to the left a view across the bay shows a series of proximal turbidites higher in the succession. Similar facies are exposed along the side of the path. The traverse can be discontinued at this point, although the coastal path continues to Banyalbufar. The essential stratigraphy, established by Rodríguez-Perea and Pomar (1983) and Jones (1984), is illustrated in Fig. 53. Again there is evidence for shallow-water deposition of limestones (Sant Elm Calcarenites) on a Triassic basement (here Buntsandstein) followed by development of an unconformity, then deepening due to major subsidence during the Burdigalian with deposition of the Banyalbufar Turbiditic Formation. The presence of chaotic mass-flow deposits in this unit is suggestive of active faulting at this time. The sediment geometry of the basal 100 metres of this clastic Miocene section is interpretable in terms of the submarine-fan model, whereas stratigraphically higher sediments, with distal turbidites, may be ascribed to a basin plain. Carbon and sulphur values of some shale samples are illustrated in Fig. 51.

Food and drink may be obtained at the Bar Restaurante Ca'n Toni Moreno: the squid and paella are recommended.

(Although not thematically appropriate to this part of the excursion it is worth noting that the only Palaeozoic basement outcrop found to date in Mallorca may

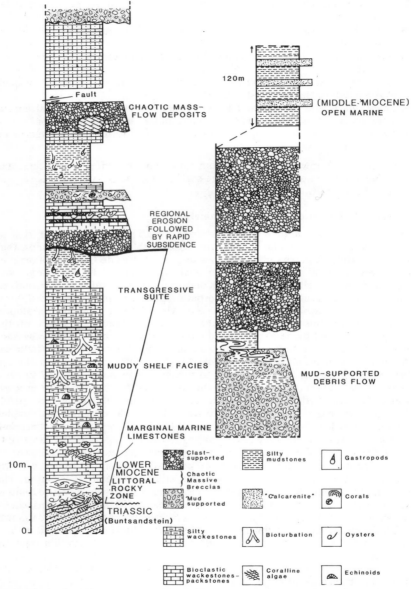

Fig. 53. Interpretative stratigraphy of the lower Miocene, Puerto de Canonge, modified after Jones (1984).

be approached from the road to Puerto de Canonge. About two kilometres down from the junction of this road with the C710 there is a dusty track that leads off (to the right, going down towards the Puerto) to descend north-eastwards. Continue until you see a sign for a quarry (Cantera) then leave the vehicles. Follow the road that descends steeply between bushes and trees to reach, after some 15 minutes walk, the sea at Es Raco de sa Cova. You will pass a house on the way. Turn right at the coast and walk along the track: the outcrop is reached in about 10 minutes. It is constituted by some 33 metres of steeply dipping black shales, siltstones and sandstones in faulted contact with Buntsandstein (Ramos and Rodríguez-Perea, 1985). These rocks show a remarkable resemblance to the Upper Devonian-Carboniferous black shales and turbidites exposed in Menorca and correlation is likely. Dating by pollen in fact shows that they are Carboniferous. Whether these Mallorcan Palaeozoic rocks are autochthonous or part of the thrust slices in the Sierra Norte is not clear. Cobbles of Palaeozoic sandstones in Tertiary sediments have been cored from a submarine rise 70km north of the coast of the Sierra Norte (Bourrouilh and Mauffret, 1975).)

Locality 19: Portals Vells

Return on the C719, direction Palma. Between Paguera and Palma Nova take the road towards Portals Vells signposted 'Cala Figuera'. Follow the road past the 'beached aeroplane' and straight on through the middle of the golf course and into the pine forest. Take the left-hand turn that leads down to the bay (labelled Portals Vells, Costa de Calvia and advertising the Bar Restaurante Es Repos) and drive to the beach (Playa Portals Vells). Walk towards the right where an artificial cave system has been dug out of the rock. The path winds over much steeply-dipping reddened Pleistocene aeolianite which is case-hardened at top, appearing more massive. Rhizocretions are common (see Locality 21 for discussion of these features). The Pleistocene locally contains included boulders.

The cave itself is constituted by large-scale cross-bedded carbonate sands containing echinoids and the trace-fossil *Ophiomorpha*. Around the cave itself the sands are particularly rich in red algae and pectinids. Cementation is patchy, but the burrows are selectively lithified. This sequence is termed (rather misleadingly as the facies are coarse-grained) the *Heterostegina* Calcisiltites Unit by Spanish geologists and it discordantly underlies the main upper Tortonian-Messinian reef facies (Pomar *et al.*, 1983b). It is dated as Tortonian.

In order to see the reef facies it is necessary to climb above the cave, where outcrops of rubbly reef rock, containing the coral *Porites*, bivalves and gastropods, with much attendant solutional porosity, are visible. Fallen blocks of this facies may also be seen in front of the cave. Looking across to the far side of the inlet at the seaward end, below a ruined watch tower, the discordant relationship between the massive reef rock and the underlying facies may also be observed. Faults affect the *Heterostegina* calcisiltites but not the overlying unit.

Returning past the cave to the opposite side of the bay, the palaeoecology of the *Heterostegina* calcisiltites may be examined. *Thalassinoides* and spiral burrows, best attributed to the action of arthropods, are particularly common in these yellowish, irregularly cemented lime sands, as are sand dollars and other echinoids. *Heterostegina, Amphistegina*, miliolids, red algae, bivalve and echinoid fragments and fish teeth can be seen in thin section. A very shallow-water, littoral environment is indicated. The facies passes laterally into conglomerates of continental character (Pomar *et al.*, 1983b).

EXCURSION 5
MESSINIAN REEF FACIES

Locality 20: Cabo Blanco (Cap Blanc)

From Palma follow the road around the bay and pick up the new highway that skirts El Arenal, then continue along the coast to Cabo Blanco: the turn-off is just after the km 18 post. Park in front of the lighthouse. Walk northwest towards the coastguard station on the cliff-top. Approximately halfway between the wall round the lighthouse and the coastguard station, a descent over the cliff-top to a lower level is possible. Pleistocene beach-dune calcarenites with rhizocretions are immediately visible. Below this a complete section through a Messinian coral reef may be seen, but the descent along a zig-zagging fishermen's path is vertiginous and not for those with no head for heights (Fig. 54). The path is usually marked with a series of paint splotches: be careful not to dislodge pebbles as people may be fishing below, particularly on week-ends.

Walk down till you encounter a brown laminated tufa-like calcareous crust: the reef section begins immediately below this. It is best to descend (with great care) to the base of the section, where the pre-reef white calcarenites, with many crustacean burrows, and locally containing *Halimeda*, are exposed. It is easy then to ascend through the various reef zones. The generalized section at Cabo Blanco is given in Fig. 55: the dominant coral genus present in the reef in this section is *Porites*, whose morphology is variable and whose preservation is generally very poor. In the lower parts of the reef the morphology is dish- or plate-like, whereas higher in the section the corals have a distinctive finger or stick shape (Fig. 56). Some of the fingers have primary dips up to 35°, presumably a response to prevailing wave and current movement. According to Pomar *et al.* (1983c, 1985) the finger *Porites* are concentrated in spurs between drainage channels (grooves), although the grooves may contain a few corals in life position. In higher levels still the corals occur in massive heads, together with red algae, bryozoans, serpulids, vermetids and echinoids; fills of *Lithophaga* bores are particularly prominent. There is considerable dolomitization. All of the originally aragonitic material, particularly *Porites*, has suffered solution producing large-scale moulds, and the corals are not all readily recognizable as such. This reef rock obviously possesses considerable primary and secondary porosity. A generalized reconstruction is given in Fig. 57, and a comparison with Recent Caribbean reefs in Fig. 58. The corals constituted a typical fringing reef around Mallorca (Fig. 59). Detailed studies of the reef by Pomar (1988) have allowed the recognition of a number of progradational-accretionary sequences of different orders of magnitude. The general accretion (first-order) is cut by important erosion surfaces (A-surfaces) associated with breccias and major facies shifts; the second-order sequences show accretional events of up to 100m with 1–2km of progradation; these are divided by less important erosion surfaces (B–surfaces) into third-order sequences

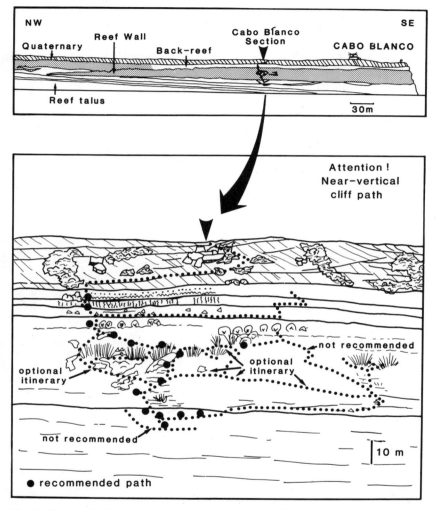

Fig. 54. Sketch of pathways descending through the Cabo Blanco section, after Pomar *et al.*, 1983c.

occurring in packages some tens of metres thick. This type of architecture implies cyclic variations of relative sea level, with rises being responsible for vertical accretion and falls for erosion and basinward progradation. The magnitude of the sea-level variations is consistent with a glacio-eustatic origin. Figs. 60 and 61 illustrate these concepts.

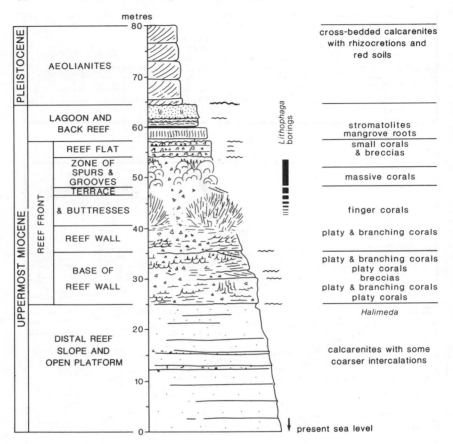

Fig. 55. Messinian reef structure of Cabo Blanco, after Pomar *et al.* (1983c).

The further significance of the Mediterranean Miocene reefs is discussed by Esteban *et al.*, (1978), Esteban (1979/80), and Jones (1984). The reefs are thought to have formed during the early part of the Messinian salinity crisis which ultimately led, after complete drawdown, to desiccation of the Mediterranean basin (Hsü *et al.*, 1978). The higher salinity tolerance of *Porites*, relative to other coral genera, may be significant in this context (cf. Kinsman, 1964). However, genera in addition to *Porites* occur in some Messinian sections on Mallorca, including Cala Pi (Locality 21).

The exuberant and diverse morphology of *Porites* is intriguing. This genus appeared in the Cretaceous and is very important in Tertiary to Recent reef environments, but only in the Messinian did it indulge in exotic growth forms. Lack of competition from other corals may provide part of the answer, in which

Fig. 56. Moulds of finger-shaped *Porites*, Messinian, Cabo Blanco.

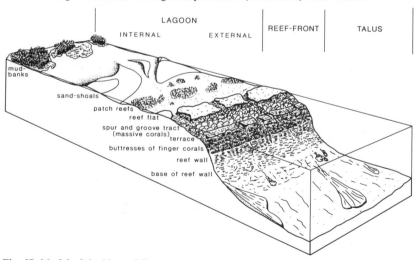

Fig. 57. Model of the Upper Miocene reef complex of Mallorca, after Pomar *et al.* (1983c).

case increased salinity was the prime mover. Esteban (1979/80) was sceptical of this, however, and suggested that the abundance of *Halimeda* and echinoids implied normal salinities. He hypothesized that the introduction of cold Atlantic waters produced physical disturbances that affected coral growth and morphology.

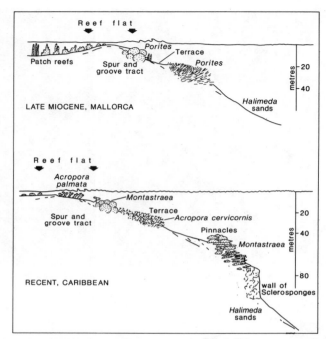

Fig. 58. Comparison between the Messinian reef of Mallorca and a Recent reef in the Caribbean, after Pomar *et al.* (1983c). Horizontal scale similar to vertical scale.

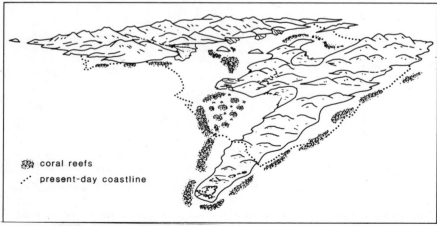

Fig. 59. Sketch of the Messinian barrier or fringing reef of Mallorca, after Pomar *et al.* (1983c).

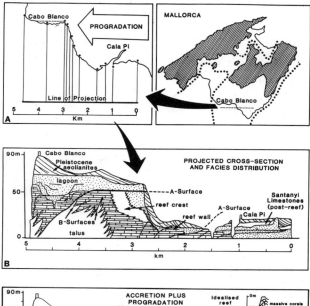

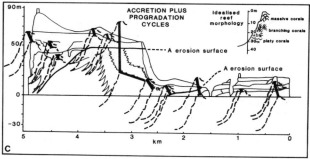

Fig. 60. Progradation geometry of the Cabo Blanco reef related to changes in relative sea level, after Pomar, 1988. A: Line of projection of the profile illustrated in B. B: Interpretation of reef architecture of the Cabo-Blanco-Cala Pi transect. Note how the reef complex is cut by a number of erosion surfaces, of which the more major (A- and B-surfaces) are illustrated here. The B-surfaces represent an erosional phase cutting down some tens of metres and their formation was associated with concurrent progradation of the reef front during sea-level fall. The A-surfaces cut through most of the reef framework and may indicate erosion up to hundreds of metres and concurrent reef advance on a scale of kilometres. Progradation was towards the south-west. C: Sketch of the reef showing progradational and aggradational cycles. Vertical migration of the idealized reef section is indicated by the bold lines: this is translated into cycles of sea-level rise as shown in the interpretative diagram in Fig. 61.

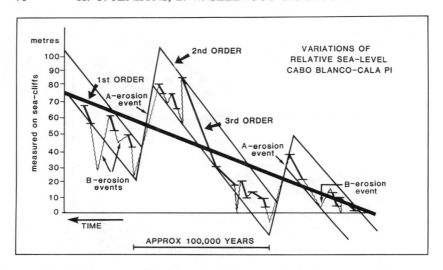

Fig. 61. Sketch to show relative sea-level variations reconstructed from the sea-cliffs in the Cabo Blanco-Cala Pi area. A three-fold hierarchy of sea-level change is recognizable, where the frequency of the B-event may correspond with the precession cycle ($\sim$21,000 years) and the A-events with the eccentricity cycle ($\sim$100,000 years) recognized in the Pleistocene record (e.g. Imbrie, 1985). Hence a glacio-eustatic model is proposed as an explanation for the architecture of the Mallorcan Miocene reefs. The bold black line indicates the general deepening trend (1st order) over a 200,000 year interval, punctuated by higher frequency cycles (2nd and 3rd order). After Pomar (1988).

Locality 21: Cala Pi

Drive to Cala Pi. Leave vehicles outside the Bar-Restaurante Miguel, facing the sea. There is a hazardous, and virtually hidden, fishermen's path that descends, at the southerly seaward extremity of the peninsula east of the ruined tower, to a prominent ledge, but this is not advised. Walk northwards parallel to the inlet, just seaward of the row of new houses, until you reach some concrete steps that descend to the sandy beach. The higher levels of the section are built of reddened Pleistocene aeolianites that are steeply dipping and full of rhizocretions.

It is perhaps useful to discuss briefly the genesis of rhizocretions at this juncture. Described in detail by Calvet *et al.* (1975), these structures are interpreted to form by progressive penetration and growth of a root producing an increased concentration of sand in its periphery. The second suggested stage is the formation of a calcitic envelope related to the activity of various micro-organisms and organic acids emanating from the living root. This first causes dissolution of the enclosing calcarenites; subsequently precipitation takes place during evaporation-transpiration with consequent changes in pH. Lastly 'chalky' material may centripetally fill some voids left by the decay of the root itself. This process is attributed to cessation of evaporation-transpiration as the plant dies and putrifaction begins, with consequent change of the microflora. A model of

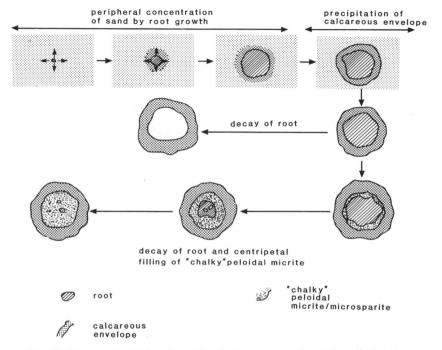

peripheral concentration
of sand by root growth

precipitation of
calcareous envelope

decay of root

decay of root and centripetal
filling of "chalky"peloidal micrite

root

"chalky"
peloidal
micrite/microsparite

calcareous
envelope

Fig. 62. Diagram illustrating the origin of rhizocretions, after Calvet *et al.* (1975).

the genesis of rhizocretions is given in Fig. 62. This evolutionary sequence was accepted in part by Klappa (1980); he believed, however, that centripetal calcareous matter may have precipitated within the living root itself. There are marked similarities, both lithological and genetic, between a vertical profile of caliche and a transverse section through a rhizocretion. A host of transitional lithological products exists between the two, and both can be considered as formed by a process of calcretization.

The contact of the Pleistocene aeolianites with the Messinian limestones is well seen a little way down the steps: the latter are constituted by shell-rich carbonates of the Santanyi Limestones (Pomar *et al.*, 1983c). In the reef-related facies below, moulds of gastropods and bivalves are very prominent, as are red algae and calcareous worm tubes. At the base of the steps there are excellently preserved corals. They may also be viewed on the far side of the inlet. The coral fauna is more diverse here than at Cabo Blanco, and Pomar *et al.*, (1983c) record, in addition to *Porites, Tarbellastraea, Siderastraea* and ?*Montastraea*. The morphology of the corals is dominantly massive and columnar. The enclosing sediment is constituted by bioturbated calcarenites with various quantities of lime mud. Both the Spanish authors and Jones (1984) interpreted the Cala Pi Messinian section as a lagoonal facies (Fig. 60).

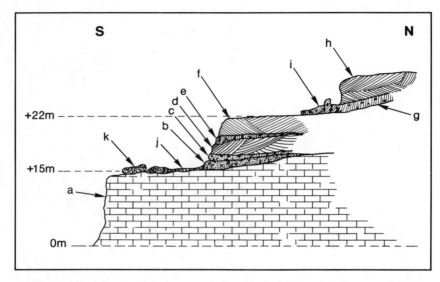

Fig. 63. Sketch of the Els Bancals section, after Cuerda (1975) and Crabtree *et al.* (1978).

The various beds present are as follows:

a) Miocene reef limestone with abundant red-algal nodules and corals: wide marine abrasion platform at 15m.

b) Sandy, reddish-yellow marls, strongly cemented and containing casts of gastropod shells and *Patella longicosta* (no longer living in the Mediterranean), perhaps of Early Pleistocene age. Some rounded pebbles or iron concretions. Thickness 20–40cm.

c) Reddish-yellow marls with a few pebbles or concretions. Thickness 30-80cm.

d) Weathered, pink, coarse-grained aeolianite with rhizocretions; 20–150cm thick.

e) Reddish marl with small angular pebbles and thin calcrete layers.

f) Grey to pink coarse-grained aeolianite with visible cross-bedding. Slightly weathered with some rhizocretions and pipes. Thickness 2–3m.

g) Pink marl with slightly rounded pebbles derived from inland. Thickness 1m or more.

h) Pink aeolianite 2–3m thick.

i) Wide abrasion platform at 22m, on which lie large blocks of (h), cemented by marly marine sands containing beach pebbles and fossils of littoral species such as *Patella*. Probably Palaeotyrrhenian.

j) Similar beach sediments on the 15m platform, also with abundant *Patella ferruginea* (now rare here) and other littoral species but none of the Senegalese species typical of the Eutyrrhenian, so this is presumably Palaeotyrrhenian.

k) Cemented on to the same platform are small patches of younger beach sediments containing such Senegalese species as *Cantharus viverratus*, typical of the Eutyrrhenian. Nearby, the same horizon also contains *Arca plicata* and *Conus testudinarius*, also typical of the same period.

Fig. 64. Well-developed red-algal nodules filling channels in Messinian reef-crest carbonates, Els Bancals. Scale in centimetres.

On the north-west side of the inlet, behind the boathouses, it is possible to climb up some steps to reach a path that runs seaward, parallel to Cala Pi. The track swings right past the head of Cala Beltran and then becomes rather indistinct among a number of other rival paths. Attempt to parallel the coast for a few hundred metres before finding a way that allows access to the cliff top. Excellent accessible exposures exist here at Els Bancals (Fig. 63) both through the Pleistocene and the underlying Messinian carbonates. The latter are locally constituted almost entirely by a unit of centimetre-scale red-algal nodules or rhodolites, interpreted as filling channels in the reef-crest (Fig. 64); these are over- and underlain by coralliferous facies: access is easy to this part of the section. The prominent platforms at 15m and 22m allow an easy passage along the cliff where the reddened Pleistocene can be examined. According to Cuerda (1975), both Lower and Upper Pleistocene sediments are present here. Crabtree *et al.* (1978) indicated three major aeolianites, each with a soil and calcrete cap and well-developed rhizocretions (Fig. 63).

Recommended stop for sustenance on return to Cala Pi: Restaurante-Bar Miguel.

Locality 22 (optional): Salinas de Levante

Drive east to the Salinas de Levante. You will need to ask permission to view the salt ponds. Note the various forms of halite present and note also the green cyanobacterial mats which are usually well developed in the absence of invertebrate algal grazers. A correlation between well-developed Phanerozoic stromatolites and ancient environments of inferred elevated salinity is commonly noted, and is relevant to understanding the algal structures seen at Locality 26 (cf. Garrett, 1970).

EXCURSION 6
PROXIMAL REDEPOSITED CARBONATES AND MESSINIAN POST-REEF OOLITES AND STROMATOLITES

Locality 23: Randa

Take the C715 from Palma and turn off to Algaida and Randa. Follow the road that leads up to the Santuario de Cura, which affords good exposures of the sequence. The structural geology of the whole area around Randa is interpreted as two thrust sheets separated by a subhorizontal thrust of mid-Miocene (Langhian) age (Anglada *et al.*, 1986): the hill itself belongs to the lower unit and its sediments are strongly cleaved. The Randa Limestone Unit is partly of equivalent age to the Banyalbufar Turbiditic Formation (Fig. 18): it is dated as Burdigalian-Langhian (Pomar *et al.*, 1983b), a time of major thrusting in the Sierra Norte. Pomar and Rodríguez-Perea (1983) have divided the Randa Limestone into three divisions: a lower turbiditic unit, an intermediate calcarenite unit, and an upper unit.

The lower turbiditic unit is exposed, albeit poorly, just outside Randa; it is constituted by marls with intercalations of calcarenites, commonly graded, laminated, locally convoluted and channellized. Included pebbles of Palaeozoic metamorphic rocks and Mesozoic limestones are common. The redeposited lime sands contain dominantly globigerinid foraminifera, red algae, bryozoans, large foraminifera such as *Amphistegina* and *Heterostegina*, and bivalves: this is a mixed planktonic-benthonic assemblage and implies the relative proximity of a carbonate shelf to the deep-water environment. Certain benthonic foraminifera in the marls have been taken to suggest depositional depths greater than 600 metres. The trace fossil *Palaeodictyon* is found. Sole marks are uncommon on the base of the lime turbidites, apart from occasional striations indicative of north-south or south-north current flow (Pomar and Rodríguez-Perea, 1983). Slumped beds are common. Lateral correlatives of this unit contain glauconite, sponge spicules and, locally, abundant siliceous phytoplankton such as diatoms, silicoflagellates, and ebrideans (Colom and Sacares, 1968; Colom, 1952, 1975b): the latter sediments are known as 'moronitas' (Fig. 17).

This turbiditic unit passes rapidly into the calcarenites above, well seen at the same spot and along the roadside about a kilometre south-east of Randa, a little before and after the turn-off to the Santuario de Gràcia. The most common structure is parallel lamination, with some beds showing grading, small-scale cross-bedding and ripple lamination. Dish and pillar structures are not uncommon. There is negligible marl in this unit and the planktonic elements in the calcarenite are vanishingly small (i.e. very few globigerines) whereas encrusting forms such as red algae and bryozoans are more important. Pomar and

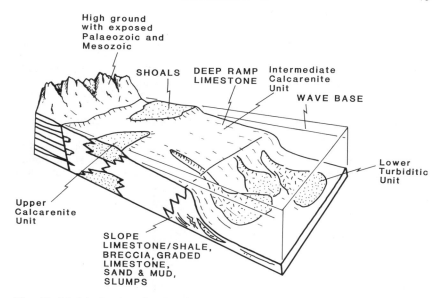

Fig. 65. Model showing the postulated peri-platform setting of the lower Miocene calcarenites of Randa, based on Pomar and Rodríguez-Perea (1983) and Read (1985). The Palaeozoic pebbles may either have been derived locally or from Menorca.

Rodríguez-Perea (1983) referred to numerous 'olistostromes' in the mid part of this unit.

Continuing along the road, toward the Santuario de Cura, for about a kilometre or so, to two or three hairpin bends below the summit, one may view the large-scale cross-bedding in the upper unit. Take a sharp turn-off to the right, labelled Sant Honorat; once arrived in the car park channellized mass-flow conglomerates and calcarenitic turbidites are spectacularly exposed. Rejoining the main road, continue to the summit, and park close to the three radio masts and flat-roofed building. Walk south-west towards the limestone cliff. Note the large-scale cross-bedded units, opposed ripples, included Palaeozoic and Mesozoic pebbles and graded units. The sequence may be interpreted as constituted by shallow-marine carbonate sand-waves. The composition of the sediments is essentially the same as in the lower unit, except that red algae and coral fragments are more common. Deposition on a ramp adjacent to a prograding carbonate platform is suggested by Pomar and Rodríguez-Perea (1983). The presence of Palaeozoic and Mesozoic lithoclasts implies rejuvenation of topography, which is consistent with the tectonic motif hypothesized for this time. The Palaeozoic pebbles have presumably been derived either from a local horst of Mallorcan basement or Menorca. A possible model is given in Fig. 65. Several other interesting outcrops exist around the summit of Randa and general exploration of the area is worthwhile.

Locality 24: Cala Llombarts

From Randa drive to Lluchmajor and follow the C717 to Santanyi and thence follow signs to Cala Llombarts (do not take the road to Cala Santanyi or Cala Figuera). As you near the coast follow signs to 'playa' where the road descends past the Restaurante Cala Llombarts to the beach. Walk towards the south-west side of the inlet where a flight of steps that leads down to beach level cuts through cross-bedded oolitic facies.

At sea level the irregular karstified surface of the reefal unit is exposed, and the overlying facies, the Santanyi Limestones, are present in the cliff behind. A generalized section is given in Fig. 66. Directly above the reefal unit, some 4 to 5 metres of limestone display a vertical rootlet system which Fornós and Pomar (1983) interpreted as produced by mangroves (Fig. 67). A possible equivalent

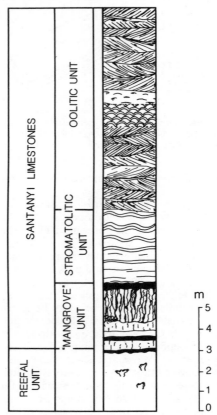

Fig. 66. Idealized section through the Messinian Santanyi Limestones exposed at Cala Llombarts. After Fornós and Pomar (1983).

Fig. 67. Peloidal limestones with root imprints interpreted as the product of a swamp colonized by mangroves by Fornós and Pomar (1983). Santanyi Limestones, Messinian, Cala Llombarts.

Fig. 68. Leached oolites showing multi-directional cross-bedding, partly of herring-bone type: deposition by tidal currents is probable. Fallen block is inverted; hence photograph is correctly oriented. Coin is 100-peseta piece: diameter 2.5cm. Santanyi Limestones, Messinian, Cala Llombarts.

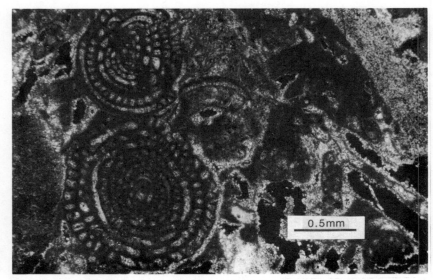

Fig. 69. Lagoonal carbonate sand rich in miliolid foraminifera. Considerable primary porosity is present. Thin section, crossed nicols, Messinian, Porto Cristo.

would be the sub-Recent mangrove reef described by Hoffmeister and Multer (1965) from Key Biscayne, Florida: horizontal as well as vertical root systems are a feature of this deposit. The sediments of the Messinian facies are peloidal micrites with abundant miliolids and ostracods. Collapse structures occur locally where underlying coral patches have dissolved. Above the 'mangrove' facies a stromatolitic level is developed, and, following that, the oolites. Fallen blocks of partly cemented oolite are visible here and can be easily studied; they show good trough and herring-bone cross-bedding (Fig. 68) but are almost entirely leached by meteoric-water diagenesis with attendant mouldic porosity. The ooids were presumably originally aragonitic in mineralogy. Some stromatolitic facies are locally developed within the oolite.

The main stromatolitic unit, intercalated between the mangrove and oolitic facies, is also well seen on the opposite side of the bay, where access to some of the section is easier. Again, a number of fallen blocks can be studied. Most of the stromatolitic micrite is parallel-laminated but some small domal structures also occur. Rippled beds occur at the base of the unit, and these are rich in the bivalve *Tellina* (Fornós and Pomar, 1983). Two separate beds of the 'mangrove' facies are visible on the northerly side of the bay if one walks into the adjacent cove. Below the lower rootlet layer there is a conglomeratic top to the reef carbonate.

Clearly all these carbonate facies are of very shallow-water character and parallels may be made with environments in the Bahamas and Persian (Arabian) Gulf. The sequence itself presumably reflects the lateral migration of facies belts

Fig. 70. Giant stromatolite, Santanyi Limestones, Messinian, Porto Pi. The diameter of the algal head is approximately 80cm.

through time in response to sediment accumulation, subsidence and changes in sea level.

Locality 25 (optional): Porto Cristo

Drive north to Porto Cristo, where the caves of Drach and Hams are situated. On a little peninsula that protrudes southwards on the east side of the harbour there is good exposure of Messinian facies, weathered with a pronounced karst topography. There is much reddening with formation of duricrusts and rhizocretions. This is a lime sand containing abundant miliolid foraminifera, red algae, echinoids, bivalves and the occasional coral clump, presumably deposited in a lagoonal environment (Fig. 69). Cementation is not complete and substantial void space remains. A spectacular collapsed cavern is present here: such phenomena are common on the south-eastern coast of Mallorca.

Locality 26: Porto Pi

Return to Palma. Porto Pi is situated on the west side of the city, its entrance close to the division of the old and new (motorway) roads to Palma Nova and Andraitx. Follow signs to Zona Portuaria and Dique del Oeste. A few hundred metres along the road, below the army barracks, the Bonanova Marls and overlying Santanyi Limestones are exposed in faulted contact (Fornós and Pomar, 1983; Pomar *et al.*, 1985). The basal unit of the Santanyi Limestones, well displayed close to the shore, is in stromatolitic facies, with the domal diameter of the algal clumps extending to 4 metres (Fig. 70). The size of the stromatolites decreases upwards and the upper surfaces show desiccation cracks. Inter-dome areas consist of ripple cross-laminated lime sands. Above, there is locally developed a 'reef' built of calcareous worm tubes; the section finishes with dolomitized oolite seen, according to Fornós and Pomar (1983), close to sea level, further along the road to Porto Pi. The sequence of facies thus shows some general resemblance to that exposed at Cala Llombarts. The presence of the giant stromatolites is suggestive of salinities elevated enough to exclude algal grazers, as occurs for example in Hamelin Pool, Shark Bay, Western Australia (Logan, 1961; Logan et al., 1974; Garrett, 1970; Walter, 1976), and they may perhaps be genetically related to the Messinian Mediterranean event. A cautionary note must, however, be sounded as giant subtidal stromatolites are presently forming in mobile oolitic sand belts, under conditions of normal salinity, on the Bahama Banks (Dravis, 1983; Dill et al., 1986).

REFERENCES

ADAMS, A. E. 1989. Genesis of fragmental carbonate rocks in the Mesozoic and Tertiary of Mallorca, Spain. *Geol. Jl*, **24**, 19–29.

ALBAIGÉS, J., J. ALGABA, E. CLAVELL & J. GRIMALT, 1986. Petroleum geochemistry of the Tarragona Basin (Spanish Mediterranean offshore). *In* (Leythauser, D. and J. Rüllkötter; eds), Advances in Geochemistry, 1985, *Org. Geochem.*, **10**, 441–450.

ALVARO-LÓPEZ, M., P. DEL OLMO ZAMORA & J. RAMIREZ DEL POZO, 1982. Baleares. *In*, El Cretácico de España, Univ. Complutense, Madrid, 633–653.

ALVARO, M., A. BARNOLAS, P. DEL OLMO, J. RAMIREZ DEL POZO & A. SIMO, 1984. Sedimentología del Jurásico de Mallorca. *In* (Barnolas-Cortinas, A.; ed.), Libro Guia de la Excursión, published for 'Grupo espanol de Mesozoico', Palma de Mallorca, 263pp.

ANGLADA, E., F. SÀBAT & P. SANTANACH, 1986. Les charriages de la zone centrale de Majorque (Baléares, Espagne): la structure de la région de Randa. *C.R. Acad. Sci. Paris, Ser. II*, **303**, 585–590.

ARBONA, J., J. M. FONTBOTÉ, GONZALEZ-DONOSO, A. LINARES, F. OLORIZ, L. POMAR, P. RIVAS & F. SÀBAT, 1984/5. Precisiones bioestratigráficas y aspectos sedimentólogicos del Jurásico-Cretácico basal de la isla de Cabrera (Baleares). *Cuad. Geol., Univ. Granada*, **12**, 169–186.

ARTHUR, M. A., W. E. DEAN, D. BOTTJER & P. A. SCHOLLE, 1984. Rhythmic bedding in Mesozoic-Cenozoic pelagic carbonate sequences: the primary and diagenetic origin of Milankovitch-like cycles. *In* (Berger, A. L., Imbrie, J., Hays, J., Kukla, G. and B. Salzman; eds.), Milankovitch and Climate, NATO ASI, Ser. C, **126**, Part I, Reidel, Dordrecht, 191–222.

AZÉMA, J., R. BOURROUILH, Y. CHAMPETIER, E. FOURCADE & Y. RANGHEARD, 1974. Rapports stratigraphiques, paléogéographiques et structuraux entre la chaîne ibérique, les cordillères bétiques et les Baléares. *Bull. Soc. géol. Fr., sér. 7*, **16**, 140–160.

AZÉMA, J., G. CHABRIER, P. CHAUVE. & E. FOURCADE, 1979. Nouvelles données stratigraphiques sur le Jurassique et le Crétacé du nord-ouest d'Ibiza (Baléares, Espagne). *Geol. Romana*, **18**, 1–21.

BAKER, P. & M. KASTNER, 1981. Constraints on the formation of dolomite. *Science*, **213**, 214–216.

BANDA, E., J. ANSORGE, M. BOLOIX & D. CÓRDOBA, 1980. Structure of the crust and upper mantle beneath the Balearic Islands (western Mediterranean). *Earth planet. Sci. Letts*, **49**, 219–230.

BARRON, E. J., M. A. ARTHUR & E. G. KAUFFMAN, 1985. Cretaceous rhythmic bedding sequences: a plausible link between orbital variations and climate. *Earth planet. Sci. Letts*, **72**, 327–340.

BARTHEL, W., F. CEDIEL, O. F. GEYER, & J. REMANE, 1966. Der Subbetische Jura von Cehegin (Provinz Murcia, Spanien). *Mitt. Bayer. Staatssamml. Paläont. hist. Geol.*, **6**, 167–211.

BATE, D. M. A. 1914a. On remains of a gigantic land tortoise (*Testudo gymnesicus* N. sp.) from the Pleistocene of Menorca. *Geol. Mag., Decade VI*, **1**, 100–107.

BATE, D. M. A. 1914b. On the Pleistocene ossiferous deposits of the Balearic Islands. *Geol. Mag., Decade VI*, **1**, 337–344.

BATE, D. M. A. 1944. Pleistocene shrews from the larger Western Mediterranean islands. *Annls Mag. Nat. Hist., Ser. II*, **11**, 738–769.

BERNER, R. A. & R. RAISWELL, 1983. Burial of organic carbon and pyrite sulphur in sediments over Phanerozoic time: a new theory. *Geochim. Cosmochim. Acta*, **47**, 855–862.

BERNOULLI, D. & H. C. JENKYNS, 1974. Alpine, Mediterrenean and central Atlantic Mesozoic facies in relation to the early evolution of the Tethys. *In* (Dott, R. H. & R. H. Shaver; eds), Modern and Ancient Geosynclinal Sedimentation. *Spec. Publ. Soc. econ. Paleontol. Mineral.*, **19**, 129–160.

BIJU-DUVAL, B., J. LETOUZEY, & L. MONTADERT, 1978. Structure and evolution of the Mediterranean Basins. In (Hsü, K. J. & L. Montadert *et al.*), *Initial Reports Deep Sea Drilling Project*, **42/1**, U.S. Government Printing Office, Washington, D.C., 951–984.

BIZON, G., J.-J. BIZON, R. BOURROUILH & D. MASSA, 1973. Présence aux îles Baléares (Méditerranée occidentale) de sediments 'messiniens' déposés dans une mer ouverte, à salinité normale. *C.R. Acad. Sci. Paris, Sér. D*, **277**, 985–988.

BIZON, G., J.-J. BIZON & B. BIJU-DUVAL, 1978. Comparison between formations drilled at DSDP Site 372 in the western Mediterranean and exposed series of land. In (Hsü, K. J. & L. Montadert *et al.*), *Initial Reports Deep Sea Drilling Project*, **42/1**, U.S. Government Printing Office, Washington, D.C., 897–901.

BOSELLINI, A. & K. J. HSÜ, 1973. Mediterranean plate tectonics and Triassic palaeogeography. *Nature*, **244**, 144–146.

BOSELLINI, A., D. MASETTI & M. SARTI, 1981. A Jurassic 'Tongue of the Ocean' infilled with oolitic sands: the Belluno Trough, Venetian Alps, Italy. *Mar. Geol.*, **44**, 59–95.

BOURGOIS, J., R. BOURROUILH, P. CHAUVE, J. DIDON, M. DURAND-DELGA, E. FOURCADE, A. FOUCAULT, J. PAQUET, Y. PEYRE & Y. RANGHEARD, 1970. Données nouvelles sur la géologie des Cordillères bétiques. *Ann. Soc. géol. Nord*, **90**, 347–393.

BOURROUILH, R. 1967. Le Dévonien de Minorque (Baléares, Espagne): Ses limites et sa place en Méditérranée occidentale. *In* (Oswald, D., ed.), Internat. Symp. on the Devonian System, **2**, Alberta Soc. Petrol. Geol., Calgary, 47–60.

BOURROUILH, R. 1970. Le problème de Minorque et des Sierras de Levante de Majorque. *Ann. Soc. géol. Nord*, **90**(4), 363–380.

BOURROUILH, R. & A. MAUFFRET, 1975. Le socle immergé des Baléares (Espagne): données nouvelles apportées par des prélèvements sous-marins, *Bull. Soc. géol. Fr., sér.* 7, **17**, 1126–1130.

BOURROUILH, R. & D. S. GORSLINE, 1979. Pre-Triassic fit and Alpine tectonics of continental blocks in the western Mediterranean. *Bull. geol. Soc. Am.*, **90**, 1074–1083.

BOUTET, C., Y. RANGHEARD, P. ROSENTHAL, H. VISSCHER & M. DURAND-DELGA, M., 1982. Découverte d'une microflore d'âge norien dans le Sierra Norte de Majorque (Baléares, Espagne). *C.R. Acad. Sci. Paris, Sér. II*, **294**, 1267–1270.

BRÉHÉRET, J. G. 1988. Episodes de sédimentation riche en matière organique dans les marnes bleues d'âge aptien et albien de la partie pélagique du bassin vocontien. *Bull. Soc. géol. Fr., Sér.* 8, **4**, 349–356.

BUSNARDO, R. 1975. Prebétique et Subbétique de Jaén à Lucena (Andalousie). Introduction et Trias. *Doc. Lab. Géol. Fac. Soc. Lyon*, **65**, 183pp.

BUTZER, K. W. 1962. Coastal geomorphology of Majorca. *Ann. Assoc. Am. Geographers*, **52**(2), 191–212.

BUTZER, K. W. 1964. Pleistocene cold-climate phenomena of the island of Mallorca. *Zeitschr. Geomorphologie*, **8**, 7–13.

BUTZER, K. W. 1975. Pleistocene littoral-sedimentary cycles of the Mediterranean Basin: a Mallorquin view. *In* (Butzer, K. W. & G. Isaac, eds), After the Australopithecines, Mouton Press, The Hague, 25–71.

BUTZER, K. W. & J. CUERDA, 1962. Coastal stratigraphy of southern Mallorca and its implications for the Pleistocene chronology of the Mediterranean Sea. *J. Geol,*, **70**, 398–416.

CALAFAT, F. 1986. Estratigrafía y Sedimentología de las litofacies del Buntsandstein de Mallorca. *In*, Abstracts 9th Congr. esp. Sedimentología, Barcelona, p. 39.

CALAFAT, F., J. J. FORNÓS, M. MARZO, E. RAMOS & A. RODRIGUEZ-PEREA, 1986/87. Icnología de vertebrados de las facies Buntsandstein de Mallorca. *Acta Geol. Hisp.*, **21/22**, 515–520.

CALVET, C., L. POMAR & M. ESTEBAN, 1975. Las rizocreciones del Pleistoceno de Mallorca. *Inst. Inv. Geol., Univ. Barcelona*, **30**, 35–60.

CLAUSS, K. 1956. Über Oberdevon-Korallen von Menorca. *Neues Jahrb. Geol. Paläont., Abh.*, **103**, (1–2), 5–27.

COLACICCHI, R., L. PASSERI & G. PIALLI, 1975. Evidences of tidal environment deposition in the Calcare Massiccio Formation (Central Apennines–Lower Lias). *In* (Ginsburg, R. N.; ed.), Tidal Deposits: a Casebook of Recent examples and Fossil Counterparts, Springer-Verlag, Berlin, 345–353.

COLOM, G. 1950. Más Allá de la Prehistoria. Una geología elemental de las Baleares. Colección Cauce, Madrid, 285pp.

COLOM, G. 1952. Aquitanian-Burdigalian diatom deposits of the North Betic strait. *J. Paleont.*, **26**, 867–885.

COLOM, G. 1955. Jurassic-Cretaceous pelagic sediments of the western Mediterranean zone and the Atlantic area. *Micropaleont.*, **1**, 109–124.

COLOM, G. 1957. *Biogeografía de las Baleares. La formación de las islas y el origen de su flora y de su fauna.* Estudio General Luliano, Palma de Mallorca, 658pp.

COLOM, G. 1965. Essai sur la biologie, la distribution géographique et stratigraphique des tintinnoïdiens fossiles. *Eclog. Geol. Helv.*, **58**, 319–334.

COLOM, G. 1967a. Sur l'interprétation des sédiments profonds de la zone géosynclinale baléare et subbétique (Espagne). *Palaeogeogr., Palaeoclimatol., Palaeoecol.*, **3**, 299–310.

COLOM, G. 1967b. Los depósitos lacustres del Burdigalense superior de Mallorca. *Mem. R. Cien. Art. Barcelona*, **38**, 327–395.

COLOM, G. 1969. Sobre la presencia del Senoniense en los lechos finales de la series geosynclinal, calizo-margosa, de Mallorca. *Bol. Soc. Hist. nat. Baleares*, **15**, 135–159.

COLOM, G. 1970. Estudio litológico micropaleontológico del Lias de la Sierra Norte y porción central de la isla de Mallorca. *Mem. R. Acad. Cien. Madrid. Ser. Cien. Nat.*, **24/2**, 87pp.

COLOM, G. 1973. Esbozo de las principales lito-facies de los depósitos jurásico-cretáceos de las Baleares y su evolución pre-orogénica. *Mem. R. Acad. Cien., Madrid, Ser. Cien. Nat.*, **25/2**, 116pp.

COLOM, G. 1975a. Las diferentes fases de contracciones alpinas en Mallorca. *Estud. Geol., Madrid*, **31**, 601–608.

COLOM, G. 1975b. *Geología de Mallorca.* 2 vols, Diputación Provincial de Baleares, Inst. Estud. Baleáricos, Gráf. Miramar, Palma, 519pp.

COLOM, G. 1976. Los depósitos continentales, Aquitanienses, de Mallorca y Menorca (Baleares). *Rev. R. Acad. Cien. Ex. Fis. Nat., Madrid*, **70/2**, 353–408.

COLOM, G. 1982. *Geomorfología de Mallorca.* Gráf. Miramar, Palma, 164pp.

COLOM, G. 1983. *Los Lagos del Oligoceno de Mallorca.* Gráf. Miramar, Palma, 166pp.

COLOM, G. 1984. *Los Foraminíferos bentónicos del Cretáceo de las Baleares; su paleontología, estratigrafía y ecología.* Gráf. Miramar, Palma. 139pp.

COLOM, G. & Y. RANGHEARD, 1966. Les couches à protoglobigérines de l'Oxfordien supérieur de l'île d'Ibiza et leurs équivalents à Majorque et dans le domaine subbétique. *Rev. Micropal.*, **9**, 29–36.

COLOM, G. & Y. RANGHEARD, 1973. Données nouvelles sur l'extension de niveaux lacustres dans les formations du Miocène inférieur de la Sierra Norte de Majorque (Baléares). *Annls sci. Univ. Besançon, sér. 3*, **18**, 115–129.

COLOM, G. & J. SACARES, 1968. Nota preliminar sobre la geología estructural de la región de Randa (Puig de Galdent)–Randa, Mallorca. *Bol. Soc. Hist. nat. Baleares*, **14**, 105–120.

COLOM, G. &. J. SACARES, 1976. Estudios sobre la geología de la región de Randa-Lluchmayor-Porreras. *Rev. Balear.*, **11**, 22−71.

COLOM, G., P. FREYTET & Y. RANGHEARD, 1973. Sur des sediments lacustres et fluviatiles stampiens de la Sierra Nord de Majorque (Baléares). *Annls sci. Univ. Besançon, sér. 3*, **20**, 167−174.

COMAS, M. C., F. OLORIZ & J. M. TAVERA, 1981. The red nodular limestones (Ammonitico Rosso) and associated facies: a key for settling slopes or swell areas in the Subbetic Upper Jurassic submarine topography (southern Spain). *In* (Farinacci, A. & S. Elmi; eds), Proc. Rosso Ammonitico Symposium, Edizioni Tecnoscienza, Rome, 113−136.

COPE, J. C. W., T. A. GETTY, M. K. HOWARTH, N. MORTON & H. S. TORRENS, 1980. A correlation of Jurassic rocks in the British Isles. Part One: Introduction and Lower Jurassic. *Spec. Rept Geol. Soc. Lond.*, **14**, 73pp.

CRABTREE, K., J. CUERDA, H. A. OSMASTON & J. ROSE, 1978. *The Quaternary of Mallorca.* Quat. Res. Assoc. Field Guide, Dec. 1978, 114p.

CUERDA, J. 1975. *Los Tiempos Cuaternarios en Baleares.* Diputación Provincial de Baleares, Inst. Estud. Baleáricos, Gráf. Miramar, Palma, 304pp.

CURZI, P., J. FORNÓS, A. MAUFFRET, R. SARTORI, J. SERRA, N. ZITELLINI, A. M. BORSETTI, M. CANALS, A. CASTELLARIN, L. POMAR, P. L. ROSSO & F. SÀBAT, 1985. The South Balearic Margin (Menorca Rise): objectives and preliminary results of the cruise BAL-84. *Rend. Soc. geol. ital.*, **8**, 91−96.

DABRIO, C. J. & D. POLO, 1985. Interpretación sedimentaria de las calizas de crinoides del Carixiense subbético. *Mediterránea, Ser. Geol., Univ. Alicante*, **4**, 55−77.

DARDER, B. & P. FALLOT, 1926. *La isla de Mallorca.* Guia de la excursión C−5 del XIV Congreso Geológico Internacional, Madrid, 125pp.

DEMAISON, G. & F. T. BOURGEOIS, 1985. Environment of deposition of Middle Miocene (Alcanar) carbonate source beds, Casablanca Field, Tarragona Basin, offshore Spain. *In* (Palàcas, J. G.; ed), Petroleum Geochemistry and Source Rock Potential of Carbonate Rocks. *Am. Assoc. Petrol. Geol. Stud. Geol.*, **18**, 151−161.

DILL, R. F., E. A. SHINN, A. T. JONES, K. KELLY & R. P. STEINEN, 1986. Giant subtidal stromatolites forming in normal salinity waters. *Nature*, **324**, 55−58.

DRAVIS, J. J. 1983. Hardened subtidal stromatolites, Bahamas. *Science*, **219**, 385−387.

DRITTENBASS, W. 1979. Sedimentologie und Geochemie von Eisen-Mangan führenden Knollen und Krusten im Jura der Trento-Zone (Östliche Südalpen, Norditalien). *Eclog. Geol. Helv.*, **72**, 313−345.

DROMART, G. & F. ATROPS, 1988. Valeur stratigraphique des biomicrofaciès pélagiques dans le Jurassique supérieur de la Téthys occidentale. *C.R. Acad. Sci. Paris*, **306**, *Sér. II*, 1365−1371.

DURAND-DELGA, M. 1981. Ouvertures océaniques de Mediterranée occidentale et dislocation des chaînes alpines. *In* (Wezel F. C.; ed.), Sedimentary Basins of Mediterranean Margins, C.N.R. Italian Prog. Oceanography, 417−431.

ELLER, M. G. 1981. The Red Chalk of eastern England: a Cretaceous analogue of Rosso Ammonitico. *In* (Farinacci, A. & S. Elmi, eds), Proc. Rosso Ammonitico Symposium, Edizioni Tecnoscienza, Rome, 207−231.

ESTEBAN, M. 1979/80. Significance of the Upper Miocene coral reefs of the Western Mediterranean. *Palaeogeogr. Palaeoclimatol. Palaeoecol.*, **29**, 169−188.

ESTEBAN, M., F. CALVET, C. DABRIO, A. BARÓN, J. GINER, L. POMAR, R. SALAS & A. PERMANYER, 1978. Aberrant features of the Messinian coral reefs, Spain. *Acta Geol. Hisp.*, **13**, 20−22.

FALLOT, P. 1922. *Étude géologique de la Sierra de Majorque.* (Thèse) Paris: Béranger, 480pp.

FALLOT, P. 1923. Le problème de l'île de Minorque. *Bull. Soc. géol. Fr., ser. 4*, **23**, 3−44.

FARROW, G. E. & G. P. DURANT, 1985. Carbonate-basaltic sediments from Cobb Seamount, northeast Pacific: zonation, bioerosion and petrology. *Mar. Geol.*, **65**, 73−102.

FISCHER, A. G. 1986. Climatic rhythms recorded in strata. In, (Wetherill, G. W., A. L. Albee & F. G. Stehli, eds), *Ann. Revs. Earth plan. Sci.*, **14**, 351–376.

FORNÓS, J. J. & L. POMAR, 1983. Mioceno superior de Mallorca: unidad Calizas de Santanyi ('Complejo Terminal'). *In,* El Terciario de la Baleares (Mallorca–Menorca), Guia de las Excursiones, 10th Congr. nac. Sedim. Menorca, Univ. Palma de Mallorca, 177–206.

FORNÓS, J. J., A. RODRÍGUEZ-PEREA & F. SÀBAT, 1984. El Mesozoico de la Serra de Son Amoixa (Serres de Llevant, Mallorca). *In,* 1st Congr. esp. Geol., Segovia, **1,** 173–185.

FORNÓS, J. J., A. RODRÍGUEZ-PEREA & J. ARBONA, 1986/87. Brechas y paleokarst en los depósitos jurásicos de la 'Serra de Tramuntana' de Mallorca. *Acta Geol. Hisp.* **21/22,** 459–468.

FORNÓS, J. J., A. RODRÍGUEZ-PEREA & F. SÀBAT, 1988. Shelf facies of the Middle-Upper Jurassic, Artà Caves ('Serres de Llevant'), Mallorca, Spain. *In,* Comm. 2nd Congr. Geol. de Espana, Granada, **1,** 75–78.

FREEMAN, T. 1972. Sedimentology and dolomitization of Muschelkalk carbonates (Triassic), Iberian Range, Spain. *Bull. Am. Assoc. Petrol. Geol.,* **56,** 434–453.

FREEMAN, T. 1975. Dispersal patterns in the Spanish Buntsandstein suggest clockwise rotation of the Balearic Block. *In,* 9th Int. Congr. int. Sedim., Nice, theme **1,** p. 10 (abstract).

GARCÍA-HERNÁNDEZ, M., J. M. GONZALEZ-DONOSO, A. LINARES, P. RIVAS & J. A. VERA, 1978. Características ambientales del Lias Inferior y medio en la Zona Subbética y su significado en la interpretación general de la Cordillera. *In,* Reunión sobre la Geodinámica de la Cordillera Betica y Mar de Alborán, 1976, Publ. Univ. Granada, 125–157.

GARCÍA-HERNÁNDEZ, M., P. RIVAS & J. A. VERA, 1979. Distributión de las calizas de llanuras de mareas en el Jurásico del Subbético y Prebético. *Cuad. Geol.,* **10,** 557–567.

GARCÍA-HERNÁNDEZ, M., A. C. LOPEZ-GARRIDO., P. RIVAS, C. SANZ DE GALDEANO, & J. A. VERA, 1980. Mesozoic palaeogeographic evolution of the external zone of the Betic Cordillera. *Geol. Mijnb.,* **59,** 155–168.

GARRETT, P. 1970. Phanerozoic stromatolites: noncompetitive ecologic restriction by grazing and burrowing animals. *Science,* **169,** 171–173.

GERMANN, K. 1971. Mangan-Eisen-führende Knollen und Krusten in jurassichen Rotkalken der Nördlichen Kalkalpen. *Neues Jb. Geol. Paläont., Mh.,* **1971,** 133–156.

GIGNOUX, M. 1950. Stratigraphic Geology. W. H. Freeman, San Francisco, 682pp.

HARDIE, L. A. 1987. Dolomitization: a critical view of some current views. *J. sedim. Petrol.,* **57,** 166–183.

HERMITE, H. 1879. *Études géologiques sur les îles Baléares, 1ère partie: Majorque et Minorque.* Paris: F. Savy, 357pp.

HOFFMEISTER, J. E. & H. G. MULTER, 1965. Fossil mangrove reef of Key Biscayne, Florida. *Bull. geol. Soc. Am.,* **76,** 845–852.

HOLLISTER, J. S. 1934. Die Stellung der Balearen in varischischen und alpinen Orogen. *Abh. Ges. Wissenschaft Göttingen, math.-phys. Kl., h.r. 3,* **10,** 121–154.

HOPPE, P. 1968. Die Geologie der Berge um Grazalema (west-Andalusien, Spanien). *Geol. Jb.,* **86,** 267–338.

HSÜ, K. J. 1977. Tectonic evolution of the Mediterranean basins in the Eastern Mediterranean. *In* (Nairn, A. E. M. & F. G. Stehli; eds), The Ocean Basins and Margins, **4A,** Plenum Press, New York, 29–75.

HSÜ K. J., L. MONTADERT, D. BERNOULLI, M. B. CITA, A. ERIKSON, R. E. GARRISON, R. B. KIDD, F. MÉLIÈRES, C. MULLER & R. WRIGHT, 1978. History of the Mediterranean salinity crisis. *In* (Hsü, K. J. & L. Montadert *et al.*),

Initial Reports Deep Sea Drilling Project, **42/1**, U.S. Government Printing Office, Washington, D.C., 1053–1078.

IMBRIE, J. 1985. A theoretical framework for the Pleistocene ice ages. *J. geol. soc. Lond.*, **142**, 417–432.

JAMES, N. P. 1984. Shallowing-upward sequences in carbonates. *In* (Walker, R. G.; ed.). Facies Models, 2nd edn., Geoscience Canada Reprint Ser., **1**, Geol. Ass. Canada, 213–228.

JEFFERIES, R. P. S. & P. MINTON, 1965. The mode of life of two Jurassic species of 'Posidonia' (Bivalvia). *Palaeontology*, **8**, 156–185.

JENKYNS, H. C. 1970. Fossil manganese nodules from the west Sicilian Jurassic. *Eclog. Geol. Helv.*, **63**, 741–774.

JENKYNS, H. C. 1971a. The genesis of condensed sequences in the Tethyan Jurassic. *Lethaia*, **4**, 327–352.

JENKYNS, H. C. 1971b. Speculations on the genesis of crinoidal limestones in the Tethyan Jurassic. *Geol. Rundsch.*, **60**, 471–488.

JENKYNS, H. C. 1974. Origin of red nodular limestones (Ammonitico Rosso, Knollenkalke) in the Mediterranean Jurassic: a diagenetic model. *In* (Hsü, K. J. & H. C. Jenkyns, eds), Pelagic Sediments: on land and under the sea. *Spec. Publ. Int. Ass. Sediment.*, **1**, 249–271.

JENKYNS, H. C. 1977. Fossil nodules. *In* (Glasby, G., ed.), Marine Manganese Deposits, Elsevier, Amsterdam, 85–108.

JENKYNS, H. C. 1980. Cretaceous anoxic events: from continents to oceans. *J. geol. Soc. Lond.*, **137**, 171–188.

JENKYNS, H. C. 1985. The early Toarcian and Cenomanian-Turonian anoxic events in Europe: comparisons and contrasts. *Geol. Rundsch.*, **74**, 505–518.

JENKYNS, H. C. 1988. The early Toarcian (Jurassic) anoxic event: stratigraphical, sedimentary and geochemical evidence. *Am. J. Sci.*, **288**, 101–151.

JONES, E. 1984. *Tertiary Deposits of Mallorca.* Unpublished Ph.D. thesis, University of Reading, 381pp.

KAUFFMAN, E. G. 1981. Ecological reappraisal of the German Posidonienschiefer (Toarcian) and the stagnant basin model. *In* (Gray, J., A. J. Boucot & W. B. N. Berry, eds), Communities of the Past, Hutchinson Ross Publishing Co., Stroudsberg, Pa, 311–381.

KENDALL, A. C. 1985. Radiaxial fibrous calcite: a reappraisal. *In* (Schneidermann, N. & P. M. Harris, eds). Carbonate Cements. *Spec. Publ. Soc. econ. Paleontol. Mineral.*, **36**, 59–77.

KINSMAN, D. J. J. 1964. Reef coral tolerance of high temperatures and salinities. *Nature*, **202**, 1280–1282.

KLAPPA, C. F. 1980. Rhizoliths in terrestrial carbonates: classification, recognition, genesis and significance. *Sedimentol.*, **27**, 613–629.

KUHRY, B. 1975. Stratigraphy of the Lower Cretaceous in the Subbetic north of Velez Blanco (Province of Almeria, S.E. Spain) with special reference to oolite turbidites. *GUA Paps Geol., Amsterdam, Ser.* 7, No. 7, 41–74.

KUHRY, B., S. W. G. DE CLERCQ & L. DEKKER, 1976. Indications of current action in Late Jurassic limestones, radiolarian limestones, *Saccocoma* Limestones and associated rocks from the Subbetic of S.E. Spain. *Sedim. Geol.*, **15**, 235-258.

LAUBSCHER H. & D. BERNOULLI, 1977. Mediterranean and Tethys. *In* (Nairn, A. E. M. & F. G. Stehli, eds), The Ocean Basins and Margins, **4A**, Plenum Press, New York, 1–28.

LEMOINE, M. 1984. Histoire mésozoïque des Alpes Occidentales: naissance et évolution d'une marge continentale passive. *In* (Boillot, G., L. Montadert, M. Lemoine & B. Biju-Duval, eds), Les Marges Continentales actuelles et fossiles autour de la France, Masson, Paris, 179–217.

LOGAN, B. W. 1961. *Cryptozoon* and associated stromatolites from the Recent, Shark Bay, western Australia. *J. Geol.*, **69**, 517–533.

LOGAN, B. W., P. HOFFMAN & C. D. GEBELEIN, 1974. Algal mats, cryptalgal fabrics, and structures, Hamelin Pool, western Australia. *In*, Evolution and diagenesis of Quaternary Carbonate Sediments, Shark Bay, western Australia. *Mem. Am. Assoc. Petrol. Geol.*, **22**, 140–194.

MACINTYRE, I. G. 1985. Submarine cements–the peloidal question. *In* (Schneidermann, N. & P. M. Harris, eds), Carbonate Cements. *Spec. Publ. Soc. econ. Paleontol. Mineral.*, **36**, 109–116.

MACHEL, H.-G. 1987. Saddle dolomite as a by-product of chemical compaction and thermochemical sulfate reduction. *Geology*, **15**, 936–940.

MARZO, M., L. POMAR, E. RAMOS & A. RODRÍGUEZ-PEREA, 1983. El Paleógeno de SW de la Sierra Norte de Mallorca. *In*, El Terciario de las Baleares (Mallorca-Menorca), Guia de las Excursiones, 10th Congr. nac. Sedim. Menorca, Univ. Palma de Mallorca, 75-90.

MAUFFRET, A. 1977. Etude géodynamique de la marge des îles Baléares. *Mem. Soc. géol. Fr., N.S.*, **132**, 96pp (1979).

MAUFFRET, A., J. AUZENDE, J. L. OLIVET & G. PAUTOT, 1972. Le bloc continental Baléare (Espagne)–extension et évolution. *Mar. Geol.* **12**, 289–300.

MÉGNIEN, C. & F. MÉGNIEN, (Co-ordinators) 1980. Synthèse géologique du Bassin de Paris, 1, Stratigraphie et Paléogéographie. *Mem. B.R.G.M.*, **101**, 466pp.

MINDSZENTY, A., A. GALÁCZ, I. DÓDONY & D. S. CRONAN, 1986. Paleoenvironmental significance of ferromanganese oxide concretions from the Hungarian Jurassic. *Chem. Erde*, **45**, 177–190.

MIŠÍK, M. 1966. Microfacies of the Mesozoic and Tertiary Limestones of the West Carpathians. Vydavatel'stvo Slovensky Akad. Vied, Bratislava, 269pp.

MOLINA, J. M., P. A. RUIZ-ORTIZ & J. A. VERA, 1984. Colonias de corales y facies oncolíticas en el Dogger de las Sierras de Cabra y Puente Genil (Subbético externo, Provincia de Córdoba). *Est. Geol., Madrid*, **40**, 455–461.

MOLINA, J. M., P. A. RUIZ-ORTIZ & J. A. VERA, 1985. Sedimentación marina somera entre sedimentos pelágicos en el Dogger del Subbético externo (Sierras de Cabra y de Puente Genil, Provincia de Córdoba). *Trab. Geol., Univ. Oviedo*, **15**, 127–146.

MORROW, D. W. & B. D. RICKETTS, 1986. Chemical controls on the precipitation of mineral analogues of dolomite: the sulfate enigma. *Geology*, **14**, 408–410.

OBRADOR, A. & J. M. FONTBOTÉ, 1983. Menorca. *In*, Libro Jubilar J.M. Rios, Geología de España, **2**, Com. nac. Geol., Inst. geol. min. Esp., 382–388.

OGG, J. G., A. H. F. ROBERTSON & L. F. JANSA, 1983. Jurassic sedimentation history of Site 534 (western North Atlantic) and of the Atlantic Tethys seaway. *In* (Sheridan, R. E. & F. M. Gradstein *et al.*), Initial Reports Deep Sea Drilling Project, **76**. U.S. Government Printing Office, Washington, 829–884.

PARÉS. J. M., F. SÀBAT & P. SANTANACH, 1986. La structure des Serres de Llevant de Majorque (Baléares, Espagne): données de la region au sud de Felanitx. *C.R. Acad. Sci. Paris, Sér. II*, **303**, 475–480.

PIALLI, G. 1971. Facies di piana cotidale nel Calcare Massiccio dell' Appennino umbro marchigiano. *Boll. Soc. geol. ital.*, **90**, 481–507.

POMAR, L. 1976. Tectónica de gravedad en los depósitos Mesozoicos, Paleógenos y Neógenos de Mallorca. *Bol. Soc. Hist. nat. Baleares*, **21**, 159–175.

POMAR, L. 1979. La evolución tectosedimentaria de las Baleares: análisis crítico. *Acta Geol. Hisp.* **14**, 293–310 (1982).

POMAR, L., 1988. Reef architecture and high-frequency relative sea-level oscillations, Upper Miocene, Spain. *In*, Abstracts 9th Reg. Meeting Int. Assoc. Sedim., Leuven, 174–175.

POMAR, L. & G. COLOM, 1977. Depósitos de flujos gravitatorios en el Burdigaliense de 'Es Racó d'es Gall-Auconassa' (Sóller, Mallorca). *Bol. Soc. Hist. nat. Baleares*, **22**, 119–136.

POMAR, L. & J. CUERDA, 1979. Los depósitos marinos pleistocénicos en Mallorca. *Acta Geol. Hisp.*, **14**, 505–513 (1982).

POMAR, L. & A. RODRÍGUEZ-PEREA, 1983. El Neógeno Inferior de Mallorca: Randa. *In,* El Terciaro de las Baleares (Mallorca–Menorca), Guia de las Excursiones, 10th Congr. nac. Sedim. Menorca, Univ. Palma de Mallorca, 115–137.

POMAR, L., M. ESTEBAN, X. LLIMONA & R. FONTARNAU, 1975. Acción de liquenes algas y hongos en la telodiagénesis de las rocas carbonatadas de la zona litoral prelitoral Catalana. *Inst. Inv. Geol., Univ. Barcelona*, **30**, 83–117.

POMAR, L., A. RODRÍGUEZ-PEREA & SANTANACH, 1983a. Rôle des charriages, des failles verticales et des glissements gravitationnels dans la structure de la Serra de Tramuntana de Mallorca (Baléares, Espagne). *C. R. Acad. Sci. Paris, Sér. II*, **297**, 607–612.

POMAR, L., M. MARZO & A. BARON, 1983b. El Terciario de Mallorca. *In,* El Terciario de las Baleares (Mallorca-Menorca), Guia de las Excursiones, 10th Congr. nac. Sedim. Menorca, Univ. Palma de Mallorca, 21–44.

POMAR, L., M. ESTEBAN, F. CALVET & A. BARON, 1983c. La unidad arrecifal del Mioceno superior de Mallorca. *In,* El Terciario de las Baleares (Mallorca-Menorca), Guia de las Excursiones, 10th Congr. nac. Sedim. Menorca, Univ. Palma de Mallorca, 139–175.

POMAR, L., J. J. FORNÓS & A. RODRÍGUEZ-PEREA, 1985. Reef and shallow carbonate facies of the Upper Miocene of Mallorca. *In* (Mila, M. D. & R. Rosell, eds), Excursion Guidebook, 6th Europ. Meeting Int. Ass. Sediment. Institut d'Estudis Ilerdencs, Lleida, Spain, 495–518.

PRESCOTT, D. M. 1988. The geochemistry and palaeoenvironmental significance of iron pisoliths and ferromanganese crusts from the Jurassic of Mallorca, Spain. *Eclog. Geol. Helv.*, **81**, 387–414.

PRESCOTT, D. M. 1989. *Mesozoic Palaeogeography of the Balearic Islands, Spain.* Unpublished D.Phil. Thesis, University of Oxford, 248pp.

RAD, U. von, 1974. Great Meteor and Josephine Seamounts (eastern North Atlantic): composition and origin of bioclastic sands, carbonate and pyroclastic rocks. *'Meteor' Forsch.-Ergebnisse, C. No.* **19**, 1–61.

RADKE, B. M. & R. L. MATHIS, 1980. On the formation and occurrence of saddle dolomite. *J. sedim. Petrol.*, **50**, 1149–1168.

RAMOS, E. & A. RODRÍGUEZ-PEREA, 1985. Découverte d'un affleurement de terrains paléozoïques dans l'île de Majorque (Baléares, Espagne). *C.R. Acad. Sci. Paris, Sér. II*, **301**, 1205–1207.

RAMOS, E., M. MARZO, L. POMAR & A. RODRÍGUEZ-PEREA, 1985. Estratigrafía y sedimentología del Paléogeno del sector occidental de la Sierra Norte de Mallorca. *Rev. Investig. geol.*, **40**, 29–63.

RAMOS-GUERRERO, E., A. RODRÍGUEZ-PEREA, F. SÀBAT & J. SERRA-KIEL, 1988. Cenozoic tectonosedimentary evolution of the Mallorca area. *In,* Abstracts Symp. Geol. Pyrenees and Betics, Barcelona, p. 6.

RANGHEARD, Y. 1971. Étude géologique des îles d'Ibiza et de Formentera (Baléares). *Mem. Inst. geol. min. Espana*, **82**, 340pp.

READ, J. F. 1985. Carbonate platform facies models. *Bull. Am. Ass. Petrol. Geol.*, **69**, 1–21.

REHAULT, J.-P., G. BOILLOT & A. MAUFFRET, 1985. The Western Mediterranean Basin. *In* (Stanley, D. J & F.-C. Wezel, eds), Geological Evolution of the Mediterranean Basin, Springer-Verlag, New York, 101–129.

RIBA, O. 1983. Las Islas Baleares en el marco geológico de la cuenca mediterránea occidental durante el Terciario. *In,* El Terciario de las Baleares (Mallorca-Menorca), Guia de las Excursiones, 10th Cong. nac. Sedim. Menorca, Univ. Palma de Mallorca, 3–20.

RODRÍGUEZ-PEREA, A. & J. J. FORNÓS, 1986. Karst deposits in the Jurassic breccias of the 'Serra de Tramuntana', Majorca. *In,* Abstracts, 7th regional meeting, Int. Ass. Sediment, Krakow, 165–166.

RODRÍGUEZ-PEREA, A. & L. POMAR, 1983. El Neógeno Inferior de Mallorca: Port d'es Canonge – Banyalbufar. *In,* El Terciario de las Baleares (Mallorca-Menorca), Guia de las Excursiones, 10th Congr. nac. Sedim. Menorca, Univ. Palma de Mallorca, 91–114.

RODRÍGUEZ-PEREA, A. & E. RAMOS, 1984. Presencia de Paleozoico en la Sierra de Tramuntana (Mallorca). *Bol. Soc. Hist. nat. Baleares,* **28,** 145–148.

RODRÍGUEZ-PEREA, A., E. RAMOS-GUERRERO, L. POMAR, X. PANIELLO, A. OBRADOR & J. MARTT, 1987. El Triásico de las Baleares. *Cuad. Geol. Ibérica,* **11,** 295–321.

SÀBAT, F., 1986. *Estructura geológica de les Serres de Llevant de Mallorca (Balears).* Ph.D. thesis, Univ. Barcelona, 2 vols., 128pp., 76 figs.

SÀBAT, F. & P. SANTANACH, 1984. Tectònica extensiva d'edat juràssica a l'illa de Cabrera (Balears). *Acta geol. Hisp.,* **19,** 227–234.

SÀBAT, F. & P. SANTANACH, 1985. Unitats estructurals de l'illa de Cabrera (Balears). *Rev. Investig. geol.,* **41,** 111–121.

SÀBAT, F., J. A. MUNOZ & P. SANTANACH, 1988. Transversal and oblique structures at the Serres de Llevant thrust belt (Mallorca Island). *Geol. Rundsch.,* **77,** 529–538.

SANDBERG, P. A. 1975. New interpretations of Great Salt Lake ooids and of ancient non-skeletal carbonate mineralogy. *Sedimentol.,* **22,** 497–537.

SANDBERG, P. A. 1983. An oscillating trend in Phanerozoic non-skeletal carbonate mineralogy. *Nature,* **305,** 19–22.

SCHLAGER, W. 1981. The paradox of drowned reefs and carbonate platforms. *Bull. geol. Soc. Am.,* **92,** 197–211.

SCHLANGER, S. O. & H. C. JENKYNS, 1976. Cretaceous oceanic anoxic events: causes and consequences. *Geol. Mijnb.* **55,** 179–184.

SCHOLLE, P. A. & M. A. ARTHUR, 1980. Carbon isotopic fluctuations in Cretaceous pelagic limestones: potential stratigraphic and petroleum exploration tool. *Bull. Am. Ass. Petrol. Geol.,* **64,** 67–87.

SCHWARZBACH, M. 1958. Die 'Tillite' von Menorca und das Problem devonischer Vereisungen. *Sonderveröffentl. Geol. Inst. Univ. Köln,* **3,** 19pp.

SEYFRIED, H. 1978. Der subbetische Jura von Murcia (Südost-Spanien). *Geol. Jb,* **29B,** 284pp.

SEYFRIED, H. 1980. Über die Bildungsbereiche mediterraner Jurasedimente am Beispiel der Betische Kordillere (Südost-Spanien). *Geol. Rundsch.,* **69,** 149–178.

STOECKINGER, W. T. 1976. Valencian Gulf: Offer deadline nears. *Oil & Gas Jl.,* March 29th, 197–204 and April 5th, 181–183.

TURNER, J. 1965. Upper Jurassic and Lower Cretaceous microfossils from the Hautes-Alpes. *Palaeontology,* **8,** 391–396.

WADSWORTH, W. J. & A. E. ADAMS, 1989. Miocene volcanic rocks from Mallorca. *Proc. Geol. Ass.,* **100,** 107–112.

WALTER, M. R. (ed.), 1976. *Stromatolites.* Devs. in Sedim., **20,** Elsevier, Amsterdam, 790pp.

WENDT, J. 1971. Genese und Fauna submariner sedimentärer Spaltenfüllungen im mediterranen Jura. *Palaeontographica, Abt.A,* **136,** 122–192.

WIEDENMAYER, F. 1963. Obere Trias bis mittlerer Lias zwischen Saltrio und Tremona (Lombardische Alpen). *Eclog. Geol. Helv.,* **56,** 529–640.

WIEDMANN, J. 1962. Unterkreide-Ammoniten von Mallorca. 1. Lieferung: Lytoceratina, Aptychi. *Akad. Wiss. Lit. Mainz, Abh. math-nat. Kl.*, no. 1, 148pp.

WINTERER, E. L. & K. HINZ, 1984. The evolution of the Mazagan continental margin: a synthesis of geophysical and geological data with results of drilling during Deep Sea Drilling Project Leg 79. *In* (Hinz, K., E. L. Winterer *et al.*), Initial Reports Deep Sea Drilling Project, **79**, U.S. Government Printing Office, Washington, 893 – 919.

WILKINSON, B. H., OWEN, R. M. & A. R. CARROLL, 1985. Submarine hydrothermal weathering, global eustasy and carbonate polymorphism in Phanerozoic marine oolites. *J. sedim. Petrol.*, **55**, 171 – 183.

INDEX